浙江省高职院校"十四五"重点立项建设教材

U0745267

单片机应用项目式教程
——基于 C51 编程+Proteus 仿真

叶 钢 主 编

张 莉 蓝天真 副主编

电子工业出版社.

Publishing House of Electronics Industry

北京·BEIJING

内 容 简 介

本书是浙江省高职院校"十四五"重点立项建设教材。本书是按照项目导向、任务驱动编写模式，以 MCS-51 系列单片机应用项目为载体，通过校企双元合作开发的集硬件电路分析、C51 程序设计和 Proteus 仿真调试于一体的项目式教材。

本书淡化原理、注重应用、软硬结合、虚实结合，将单片机应用设计与开发所必需的基础理论知识与实际技能分解到不同项目和任务中。本书结构紧凑、图文并茂，程序编写思路简洁清晰，易于学生理解。本书配套提供 100 多个二维码，可供读者获取各个任务的微课视频、任务仿真效果视频、导学材料、参考程序等学习资源，从而可自行完成多个任务的设计制作，达到学以致用的效果。

本书由中高职一体化教学团队共同编写，可作为中等职业院校、高等职业院校电子信息类、自动化类、计算机类等相关专业的中高职一体化教材，也可供单片机初学者、电子爱好者等相关人员学习和参考。

图书在版编目（CIP）数据

单片机应用项目式教程：基于 C51 编程+Proteus 仿真 /
叶钢主编. -- 北京 ：电子工业出版社，2025. 5.
ISBN 978-7-121-50390-0

Ⅰ. TP368.1

中国国家版本馆 CIP 数据核字第 2025BZ1234 号

责任编辑：王艳萍
印　　刷：三河市双峰印刷装订有限公司
装　　订：三河市双峰印刷装订有限公司
出版发行：电子工业出版社
　　　　　北京市海淀区万寿路 173 信箱　　邮编：100036
开　　本：787×1 092　1/16　印张：15.25　字数：390.4 千字
版　　次：2025 年 5 月第 1 版
印　　次：2025 年 5 月第 1 次印刷
定　　价：55.00 元

凡所购买电子工业出版社图书有缺损问题，请向购买书店调换。若书店售缺，请与本社发行部联系，联系及邮购电话：（010）88254888，88258888。

质量投诉请发邮件至 zlts@phei.com.cn，盗版侵权举报请发邮件至 dbqq@phei.com.cn。

本书咨询联系方式：（010）88254574，wangyp@phei.com.cn。

　　本书是在编者二十多年单片机应用开发和教学改革经验的基础之上，紧扣"知行合一、工学结合""教学做一体"的职业教育办学理念，按照项目导向、任务驱动编写模式，以MCS-51 系列单片机应用项目为载体的融合单片机硬件电路分析、C51 编程和 Proteus 仿真调试于一体的项目式教材。

　　本书在编写的过程中主要有以下特色与创新。

　　（1）项目引领，任务驱动。

　　本书内容由易到难，由浅入深，循序渐进，以"学习目标""设计要求""任务实施"贯穿每个任务，并在每个项目后加入"素养小课堂""课后任务""知识拓展"模块，让学生了解、体验单片机应用的乐趣，强化基础学习的体验感，扩展学生的专业视野，内化形成良好的职业素养，提升学生的实践能力。

　　（2）采用 C 语言编程，贴近职业岗位的需求。

　　本书选用 C 语言作为编程语言。由于 C 语言编程的难度相对较低，开发速度快，可移植性好，因此其成为企业进行单片机应用系统开发的主流语言。目前，基于单片机的应用开发，企业工程师在没有特殊要求的情况下，一般采用 C 语言编程，从而更贴近职业岗位的需求。

　　（3）校企合作，双元开发，模块化编程，对接工程实际。

　　本书编者既有来自高校的教师，又有来自企业一线的单片机工程师，实际工程经验非常丰富。本书介绍了"模块化编程"思路，这是现代企业绝大多数工程师在程序开发过程中采用的编程方法。本书的内容很多都源于实际产品的设计制作，无论是元器件的选型、电路的设计，还是程序的编写，都反映了工程上的实际需求，注重融入了现代企业的新技术、新工艺、新规范。

　　（4）借助单片机仿真软件 Proteus，融"教、学、做"于一体。

　　本书力求做到通俗性、可读性、阶梯性和实用性，借助单片机仿真软件 Proteus，融"教、学、做"于一体，使抽象的原理变得生动易学，便于教师的教学工作，强化学生动手能力的培养，也便于单片机初学者的学习。

　　（5）配套丰富的新形态教学资源。

　　本书配套丰富的新形态教学资源，通过扫描二维码可观看微课视频、导学材料、源程序等数字化教学资源，适合教师、学生随扫随学，助教助学。

　　（6）课程思政融入。

　　本书每个项目针对不同知识的特点及需求，融入思政元素。本书在每个任务中都列写了"思政目标"，并在每个项目后配有素养小课堂，让学生在掌握专业知识技能的同时，培养学生的思想政治素质和道德素质，达到既教书又育人的效果，实现"全课程育人"的功能。

本书将单片机教学内容分为 10 个项目，每个项目包含 2 个任务，从而将 C51 的相关知识点打散融入每个具体的任务进行教学。每个任务基本都包括以下内容：设计要求、设计原理（理论知识点介绍）、硬件电路设计、程序设计、系统仿真等，完全按照工作过程项目化进行编写工作。学生通过自学或给予适当的指导，均可以独立完成产品设计。这种教学方式具有体系结构新颖、知识综合运用性强、理论紧密联系实际、能够启发思考、易于自学等特点。

本书是浙江省高职院校"十四五"重点立项建设教材，由校企双元合作开发和编写。本书由丽水职业技术学院叶钢老师担任主编，丽水职业技术学院张莉老师、浙江聚效电子科技有限公司蓝天真工程师担任副主编。其中，叶钢老师负责全书内容的组织、统稿，以及项目一、项目五、项目六、项目七、项目九、项目十的内容编写，张莉老师负责项目二、项目三、项目四的内容编写，蓝天真工程师负责项目八的内容编写。本书配套的微课视频由丽水职业技术学院叶钢、张莉、张景森、李如意四位老师主讲。

由于编者水平所限，书中难免存在疏漏及不足之处，敬请读者批评指正。

编　者

目 录

项目一　简易信号灯系统的仿真设计

众所周知，近几十年来微型计算机的发展十分迅速，其发展方向主要有两个：其一是不断推出高性能的通用微型计算机系统；其二是面向控制型应用领域的单片微型计算机（单片机）的大量生产和广泛应用。由于单片机具有可靠性高、体积小、价格低、易于产品化等特点，因而在智能仪器仪表、实时工业控制、智能终端、通信设备、导航系统、家用电器等领域获得了广泛应用。

什么是单片机
微课视频

任务 1.1　闪烁灯的仿真设计

学习目标

【知识目标】

（1）了解单片机的定义。

（2）掌握单片机的常用封装形式。

（3）认识单片机各引脚及其功能。

（4）掌握振荡电路和复位电路的组成。

（5）构建单片机最小系统电路。

（6）了解并掌握单片机 I/O 接口的使用方法。

（7）了解并掌握 C51 基础知识。

闪烁灯导学材料

【技能目标】

（1）了解并掌握单片机仿真软件 Proteus 的使用方法。

（2）了解并掌握单片机编译软件 Keil C51 的使用方法。

（3）了解并掌握单片机程序下载的方法。

（4）了解并掌握单片机最小系统的组成。

（5）通过闪烁灯的仿真设计初步了解并掌握单片机应用项目的开发步骤。

【思政目标】

（1）了解单片机从业人员应当具备的职业道德守则，为从事电子信息行业做好准备。

（2）理解并敬重工匠精神，在单片机程序设计中，培养学生态度认真、一丝不苟的工匠精神。

（3）通过介绍国内外芯片的发展趋势，激发学生的爱国情怀、社会责任感和使命感。

1.1.1　单片机的基础知识

1. 定义

由于单片机的结构及功能均按工业控制要求设计，因此其确切的名称应是单片微控制器。

单片机将中央处理器（CPU）、随机存取存储器（RAM）、只读存储器（ROM）、I/O 接口电路、定时器/计数器及串行接口等集成在一块芯片上，构成一个完整的微型计算机，故又称单片微型计算机。

2. 封装形式

Intel 公司于 1980 年推出了 MCS-51 系列单片机，它是一种高性能的 8 位单片机，典型产品为 8051。

MCS-51 系列单片机有 2 种封装形式，一种是采用 40 只引脚的双列直插式封装（DIP），另一种是方形封装，如图 1-1 所示。其中，方形封装的单片机有 44 只引脚，但其中 4 只引脚（标有 NC 的引脚 1、12、23、34）是不使用的。不管是 DIP 还是方形封装，40 只引脚都可分为 3 个部分：4 个并行接口共有 32 只引脚，可分别用作地址线、数据线和 I/O 接口线；6 只引脚用作控制信号线；2 只引脚用作电源线。

MCS-51 系列单片机的内部集成如下。

（1）8 位 CPU。

（2）4KB 的片内程序存储器（片内 ROM）。

（3）128B 的片内数据存储器（片内 RAM）。

（4）64KB 的片外程序存储器（片外 ROM）的寻址能力。

（5）64KB 的片外数据存储器（片外 RAM）的寻址能力。

（6）32 根 I/O 接口线。

（7）1 个全双工异步串行接口。

（8）2 个 16 位定时器/计数器。

（9）5 个中断源，2 个优先级。

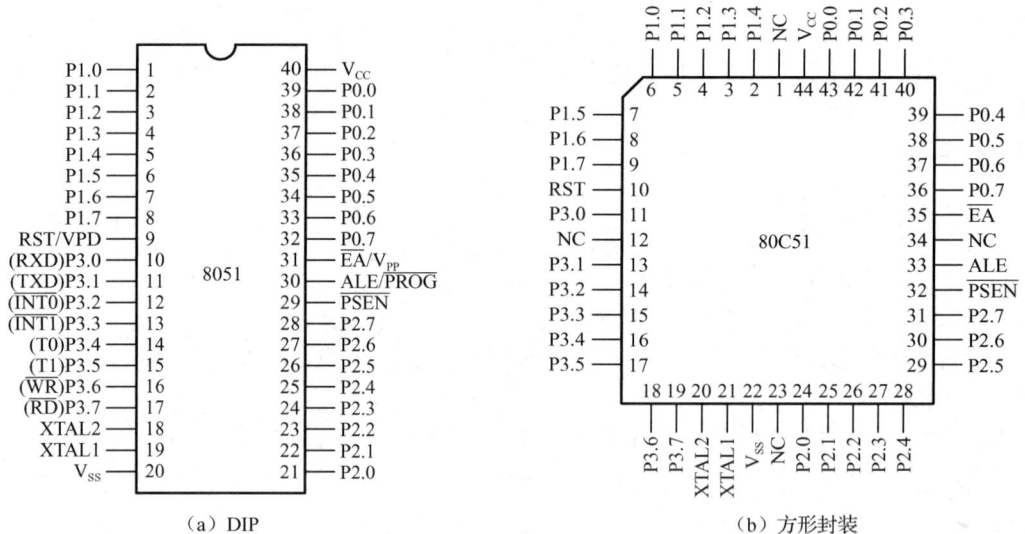

（a）DIP   （b）方形封装

图 1-1　MCS-51 系列单片机的封装形式

3. AT89 系列单片机

Atmel 公司在 8051 的基础上推出了 8 位 Flash 单片机——AT89 系列单片机。它以 MCS-51 系列单片机为内核，与 MCS-51 系列单片机的软硬件兼容。AT89 系列单片机有着十分广泛的用途，在便携式、省电、保存特殊信息的仪器和系统中显得更为有用。

AT89 系列单片机主要有 7 种型号，分别为 AT89C51、AT89LV51、AT89C52、AT89LV52、AT89C2051、AT89C1051、AT89S8252。其中，AT89C51、AT89LV52 分别是 AT89LV51、AT89C52 的低电压产品，最低电压为 2.7V。而 AT89C2051、AT89C1051 则是低挡型的低电压产品，它们只有 20 只引脚，最低电压也为 2.7V，如表 1-1 所示。

表 1-1　AT89 系列单片机常用产品特性一览表

型号	AT89C51	AT89C52	AT89C1051	AT89C2051	AT89S8252
Flash/KB	4	8	1	2	8
片内 RAM/B	128	256	64	128	256
I/O 接口线	32	32	15	15	32
定时器/计数器	2	3	1	2	3
中断源	6	8	3	6	9
串行接口	1	1	1	1	1
EEPROM/KB	无	无	无	无	2

4. 振荡电路

振荡电路就是在 AT89C51 的引脚 18（XTAL2）、引脚 19（XTAL1）两端串联一个 12MHz 的晶体振荡器，在晶体振荡器两端各接一个 30pF 的瓷片电容到地，如图 1-2 所示。两个电容一般取 30pF 左右，而晶体振荡器的频率范围通常是 1.2MHz～24MHz，晶体振荡器的频率越高，振荡频率就越高。

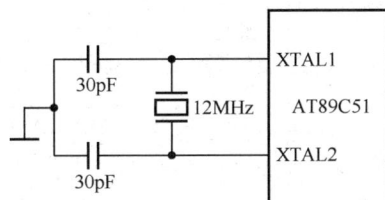

图 1-2　振荡电路

5. 复位电路

AT89C51 的引脚 9（RST）为复位引脚，当晶体振荡器工作时，在此引脚处持续给出两个机器周期的高电平可以完成复位（为了保证应用系统可靠复位，通常使 RST 引脚保持 10ms 以上的高电平）。根据这个原则，通常采用以下几种复位电路。

（1）上电自动复位电路。

如图 1-3（a）所示，只要电源 V_{CC} 的上升时间不超过 1ms，就可以实现上电自动复位，接通电源即可完成系统的复位初始化。

（2）按键电平复位电路。

按键电平复位是通过使复位端经电阻与电源 V_{CC} 接通实现的，其电路如图 1-3（b）所示。

（3）按键脉冲复位电路。

按键脉冲复位是利用 RC 微分电路产生的正脉冲实现的，其电路如图 1-3（c）所示。

本书一般采用如图 1-3（b）所示的按键电平复位电路。

6. I/O 接口

AT89C51 共有 4 组双向的 8 位 I/O 接口（P0 口～P3 口），每组 I/O 接口既可以一起使用，又可分开使用。但是在使用时要注意，P0 口作为输出接口时需要外接上拉电阻，P3 口除了具有输入/输出功能，它的每位接口线还具有第二功能，如表 1-2 所示。

（a）上电自动复位电路　　　（b）按键电平复位电路　　　（c）按键脉冲复位电路

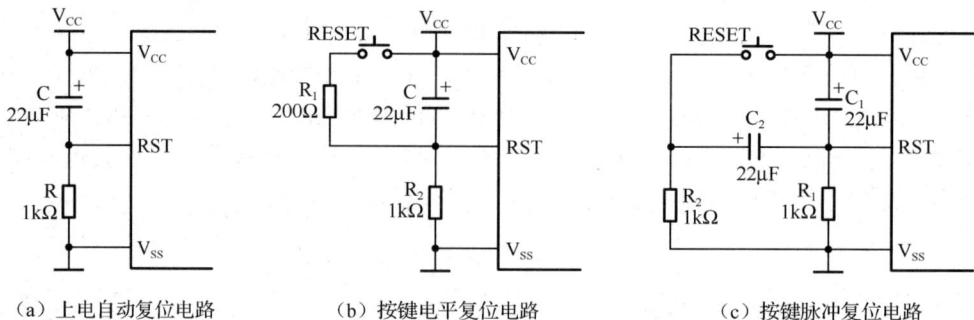

图 1-3　复位电路

表 1-2　P3 口第二功能表

引脚	功能	引脚	功能
P3.0	RXD（串行输入通道）	P3.4	T0（定时器/计数器 0 外部输入）
P3.1	TXD（串行输出通道）	P3.5	T1（定时器/计数器 1 外部输入）
P3.2	$\overline{\text{INT0}}$（外部中断 0）	P3.6	$\overline{\text{WR}}$（外数据存储器写选通）
P3.3	$\overline{\text{INT1}}$（外部中断 1）	P3.7	$\overline{\text{RD}}$（外数据存储器读选通）

1.1.2　C51 的基础知识一

C 语言在工业、计算机等方面得到了广泛应用，很多硬件开发都使用 C 语言进行编程，如 DSP、单片机、ARM 等。由于 C 语言程序本身不依赖机器的硬件系统，减轻了开发者对硬件系统的依赖，因此 C 语言程序可以不做修改或做简单的修改从不同的系统中移植过来直接使用，增加了程序的可读性和可维护性。

什么是C51 程序
微课视频

C 语言比汇编语言在编程方面有更强的优势：不需要掌握单片机汇编指令就可以直接用 C 语言进行编程；寄存器的分配将由编译器自动管理；程序更加结构化；C 语言库中有很多标准函数，数据处理能力比汇编语言的强；具有方便的模块化编程技术，使得编好的程序很容易移植。

1. C51 的基本结构

C51 的基本结构一般包含 4 个部分：头文件、声明区、主程序和函数，每部分都有一定的功能。在这些表述中，有些词语（如函数名或变量名）可以根据使用的需要自己定义，如 LED（变量）、display（函数）；有些词语是其他用户为了方便使用，定义好之后再供我们使用的，如 reg52.h 头文件中的词语；有些词语是 C 语言本身就具备的保留字，如 int、for 等，常用的保留字是需要人们记住的。

（1）头文件。

头文件是一种预先定义好的基本数据或函数。在 51 单片机中，reg51.h 或 reg52.h 头文件是定义好的内部寄存器地址的数据。用户也可以自己定义函数，然后加入头文件。

reg52.h 头文件

reg52.h 头文件将单片机常用的资源全部用大家熟悉的字符进行了定义，这样用户可以不用像汇编语言那样关注硬件底层，而只需关心控制任务。例如，在主函数中，用户写出"P3=0x01"等语句，只要程序中添加了#include<reg52.h>，编译器就会自动认出 P3 这个字符表示单片机硬件中 0xb0 这一地址单元的缓存，从而将数据 0x01 送入 0xb0 这一地址单元。

在程序中添加头文件有两种书写方法，分别为#include<reg51.h>和#include"reg51.h"。使用< >包含头文件时，编译器进入 C:\keil\C51\INC 这个文件夹（默认路径，如果 Keil 软件没有装在 C 盘，则路径不同）查找，找不到就报错；使用" "包含头文件时，编译器先进入当前工程所在文件夹查找，找不到就报错。

（2）声明区。

在指定头文件之后，可声明程序中所使用的常数、变量、函数等，其作用域将扩展至整个程序，包括主程序与所有函数。

函数可以放置在主程序之前或之后，但是函数使用之前必须预先声明。当函数放置在主程序之前时，函数的声明和定义同时完成；当函数放置在主程序之后时，必须在主程序之前进行声明，在主程序之后进行定义。从程序的简洁方面来看，函数放置在主程序之前比较好，这时函数的声明和定义同时完成，在程序中调用这些函数也很自然，程序的可读性也比较好。

（3）主程序。

主程序又称主函数，以 main()开头。主程序中分为声明区和程序区，在声明区内声明的常数、变量等仅适用于主程序，而不影响其他函数。当然，主程序也可以在声明区中定义变量，不同的是前者是局部变量，只在某个区域有效，后者是全局变量，在全程序范围内都可以使用。

（4）函数。

函数是一种具有独立功能的程序，其结构与主程序的结构类似，不过，函数可将所要处理的数据传入函数本身，称为形式参数；也可将处理完成后的结果返回调用该函数的程序，称为返回值。不管是形式参数还是返回值，在定义函数的第一行中都应该交代清楚，其通用格式如下：

```
返回值数据类型    函数名(数据类型    形式参数)
```

例如，将一个无符号字符型参数（unsigned char t）传入函数，执行完毕后返回一个整型（int）值。若函数名为 delay，则该函数对照上述通用格式的写法如下：

```
int delay(unsigned char t)
```

若不需要传入函数，则可在小括号中指定为 void 或为空。同样，若不需要返回值，则可在函数名左边指定为 void。另外，函数的内容结构形式与主程序的内容结构形式一样。

在 C51 程序中可以使用多个函数，并且函数中也可以调用函数。

（5）注释。

注释其实就是对程序进行相应的说明，C51 程序的注释一般有两种，一种是段落注释，以"/*"开始，以"*/"结束，可以对多行程序一起进行注释；另一种是行注释，以"//"开始，仅对当前行的程序进行注释。

2. 数据与数据类型

数据：具有一定格式的数字或数值。

数据类型：数据的不同格式。

常见的 C51 数据类型如表 1-3 所示。

表 1-3　常见的 C51 数据类型

数据类型	长度/bit	长度/B	范围
bit	1		0 或 1（一般用于定义位变量）
sbit	1		0 或 1（一般用于定义 I/O 接口）
char	8	1	−128～+127
unsigned char	8	1	0～255
int	16	2	−32768～+32767
unsigned int	16	2	0～65535
long	32	4	-2^{31}～$+2^{31}$
unsigned long	32	4	0～$2^{32}-1$

3. 赋值运算符与赋值表达式

在 C51 语言中，符号"="称为赋值运算符。由赋值运算符组成的表达式称为赋值表达式，其一般形式为：

变量=表达式

赋值运算的功能：先求出赋值运算符右边表达式的值，然后将该值赋给赋值运算符左边的变量，确切地说，是把数据放入以该变量为标识的存储单元中。在程序中，可以多次给一个变量赋值，因为每赋值一次，与它对应的存储单元中的数据就被更新一次。

在使用赋值运算符时，应注意以下几点。

① 赋值运算符与数学中的等号是不同的，其含义不是等同的关系，而是"赋予"的关系。例如：

i=i+1

是合法的赋值表达式，表示将变量 i 的值加 1 后重新放回到变量 i 中进行保存。

② 赋值运算符的左边只能是变量，不能是常量或表达式。例如：

5=a，a+b=c

都不是合法的赋值表达式。

③ 赋值运算符的右边也可以是一个合法的赋值表达式。例如：

a=b=c+1

是合法的赋值表达式，表示将变量 c 的值加 1 后赋给变量 a 和变量 b。

4. 自增减运算符

自增减运算符的作用是使变量值自动加 1 或减 1。例如：

++i（--i）：在使用变量 i 之前，使 i 加（减）1。

i++（i--）：在使用变量 i 之后，使 i 加（减）1。

粗略来看，++i 和 i++的作用都相当于 i=i+1，但是++i 和 i++的不同之处在于++i 先执行 i=i+1，再使用变量 i 的值；而 i++则是先使用变量 i 的值，再执行 i=i+1。

例如，已知变量 i 的值为 5，则

j=++i：执行语句后 j 的值变为 6，而 i 的值也变为 6。

j=i++：执行语句后 j 的值变为 5，而 i 的值变为 6。

注意：

（1）自增运算符（++）和自减运算符（--）只能用于变量，不能用于常量表达式。

（2）自增运算符（++）和自减运算符（--）的结合方向是"自右向左"。

5. 顺序结构与基本语句

作为结构化程序设计的一种，C51 语言同样具有顺序、分支、循环三种基本结构，并提供了丰富的可执行语句形式来实现这三种基本结构。

基本语句主要用于顺序结构程序的编写，包括赋值语句、函数调用语句、复合语句、空语句等。在 C51 语言中，语句的结束符为分号";"。

（1）赋值语句。

在任何合法的赋值表达式的尾部加上一个分号";"，就构成了赋值语句，其一般形式如下：

```
变量=表达式;
```

例如：

```
a=a+b        赋值表达式
a=a+b;       赋值语句
```

赋值语句的作用是先计算赋值运算符右边表达式的值，然后将该值赋给赋值运算符左边的变量。

赋值语句是一种可执行语句，应当出现在函数的可执行部分。

（2）函数调用语句。

在 C51 语言中，若函数仅进行某些操作而不返回值，则函数的调用可作为一条独立的语句，称为函数调用语句，其一般形式如下：

```
函数名 (实际参数表);
```

例如：

```
delayms(100);//调用 100ms 延时函数
```

或

```
display();//调用显示函数
```

（3）复合语句。

在 C51 语言中，把多条语句用一对大括号"{}"括起来组成的语句称为复合语句，又称"语句块"，其一般形式如下：

```
{
    语句1;
    语句2;
    …;
```

```
        语句 n;
    }
```

注意："{}"之后不能加";"。

复合语句虽然由多条语句组成，但它是一个整体，其作用相当于一条语句，凡可以使用单一语句的位置都可以使用复合语句。在复合语句中，不仅可以有执行语句，还可以有变量定义（或说明）语句。

C51while 语句的使用微课视频

6. while 语句

while 语句是一种循环结构的语句，其一般形式如下：

```
while(表达式)
{
    语句;//循环体
}
```

其中，表达式是 while 循环能否继续的条件，而语句则是循环体，是执行重复操作的部分。只要表达式为真，就可以重复执行循环体内的语句；反之，则退出 while 循环，执行循环体外的下一行语句。while 语句的执行过程如图 1-4 所示。while 语句的特点是先判断后执行。

7. do while 语句

do while 语句也是一种循环结构的语句，其一般形式如下：

```
do
{语句;}    //循环体
while(表达式);
```

其中，表达式可以是 C51 语言中任意合法的赋值表达式，其作用是控制循环体是否执行；循环体可以是 C51 语言中任意合法的可执行语句；最后的";"不能省略，它表示 do while 语句的结束。do while 语句的执行过程如图 1-5 所示。do while 语句的特点是先执行后判断。

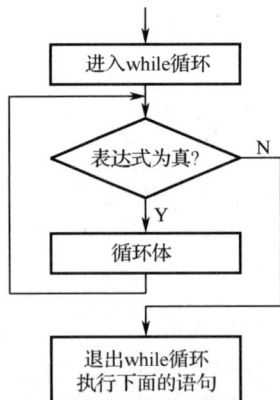

图 1-4 while 语句的执行过程

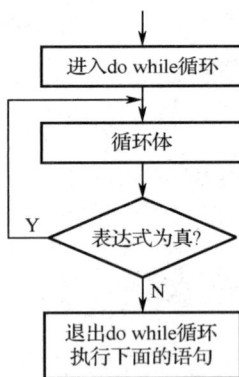

图 1-5 do while 语句的执行过程

1.1.3 闪烁灯的任务实施

【设计要求】

通过单片机驱动一个发光二极管，实现闪烁灯的效果，要求发光二极管不停地一亮一灭，时间间隔为 0.2s，循环往复。

【任务分析】

单片机驱动发光二极管可以采用两种方式进行：高电平驱动和低电平驱动。由于 51 单片机的拉电流较小（电流从 I/O 接口流出称为拉电流），灌电流较大（电流从 I/O 接口流入称为灌电流），因此一般单片机驱动发光二极管都采用低电平驱动。

如图 1-6 所示，P0.0 口通过限流电阻与发光二极管的阴极连接，发光二极管的阳极接电源，这样就实现了发光二极管的驱动电路。

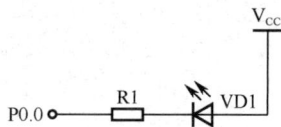

图 1-6　发光二极管的驱动电路

【实施步骤】

1. 创建仿真系统

（1）打开仿真软件 Proteus。

Proteus 仿真软件使用视频

双击仿真软件 Proteus 的图标，即可进入 Proteus ISIS 软件，可以看到如图 1-7 所示的软件界面。与其他常用的软件一样，Proteus 软件设有菜单栏、可以快速执行命令的按钮工具栏和各种各样的窗口（如原理图编辑窗口、原理图预览窗口、对象选择窗口等）。

（2）添加仿真元器件。

通过以下两种方法可以弹出元器件库选择对话框。

① 单击对象选择窗口上的"P"按钮 P L　DEVICES 。

② 按下快捷键"P"（在英文输入法下）。

若要用电阻，则可以选择 RES 库。从元器件库中选择"RES"（可以在"Keywords"处输入"RES"），在原理图预览窗口中可以看到所选择的元器件（见图 1-8），在库列表中双击该元器件或单击"确定"按钮（也可以按下"Enter"键）即可选择该电阻，同时其出现在对象选择窗口中（见图 1-8）。

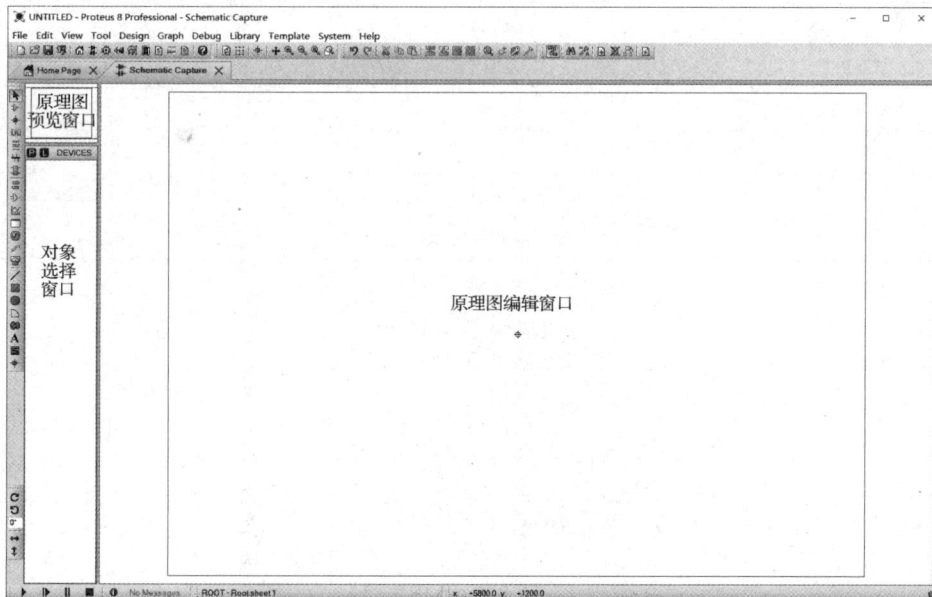

图 1-7　Proteus ISIS 的软件界面

图 1-8　元器件库选择对话框

　　按照上述的元器件添加方式，完成表 1-4 所示的元器件添加，添加元器件后的效果如图 1-9 所示。

表 1-4　仿真元器件清单

元器件名称	所属类	所属子类
AT89C51	Microprocessor ICs	8051 Family
CRYSTAL	Miscellaneous	—
CAP	Capacitors	Generic
CAP-ELEC	Capacitors	Generic
RES	Resistors	Generic
BUTTON	Switches & Relays	Switches
LED-RED	Optoelectronics	LEDs

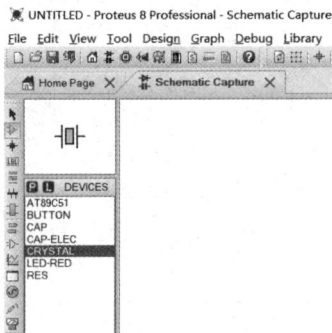

图 1-9　添加元器件后的效果

（3）绘制仿真图。

元器件全部添加后，在原理图编辑窗口中按图1-10绘制闪烁灯仿真图。

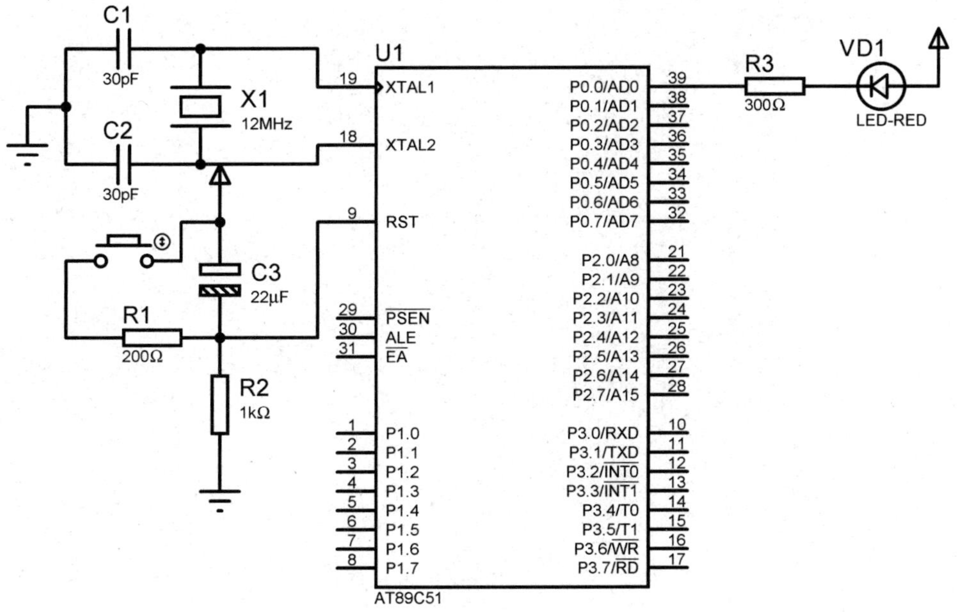

图1-10　闪烁灯仿真图

2. 编写程序

系统程序主要包括主程序和延时函数两部分，其中延时函数采用 2 个 while 语句实现200ms 延时，主程序通过对 P0.0 口进行置位和复位操作实现发光二极管的闪烁功能。主程序流程图如图1-11所示。

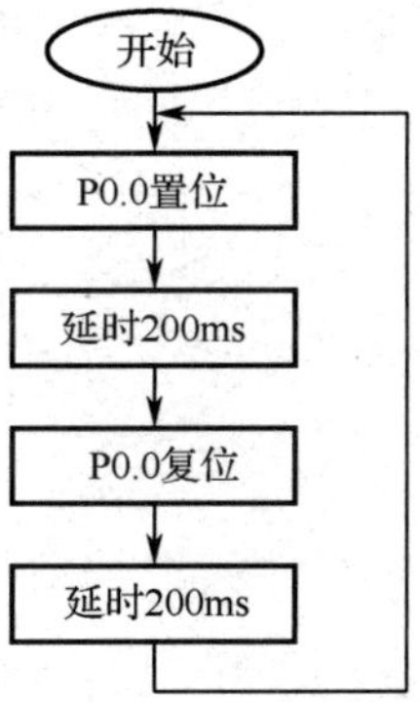

图1-11　主程序流程图

Keil C51 的使用视频

（1）打开单片机编译软件。

双击单片机编译软件 Keil μVision5 的图标，即可进入 Keil μVision5 软件，可以看到如图1-12所示的软件界面。

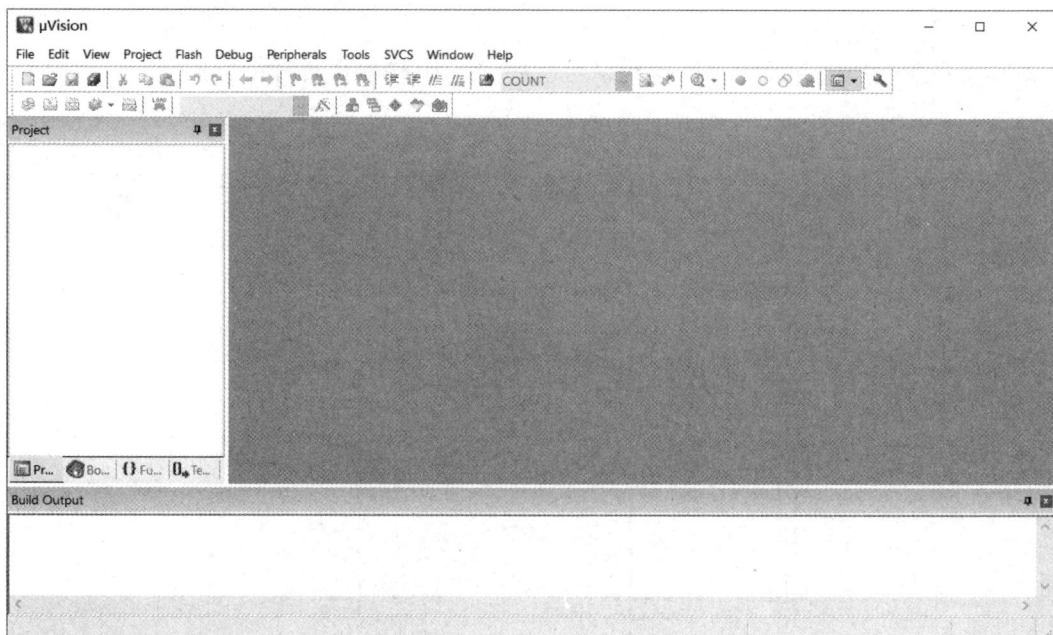

图 1-12 Keil μVision5 的软件界面

（2）新建项目。

在 Keil μVision5 的菜单栏中执行"Project"→"New μVision Project…"命令，弹出"Create New Project"对话框，选择目标路径，在"文件名"输入框中输入项目名，如图 1-13 所示。

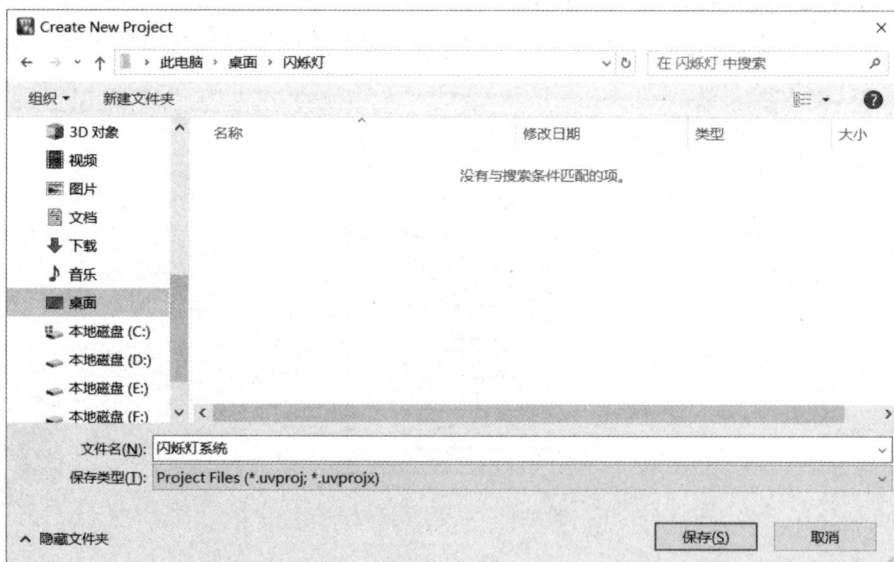

图 1-13 "Create New Project"对话框

单击"保存"按钮，弹出"Select Device for Target 'Target 1'"对话框。在此对话框的"Search"输入框中输入"AT89C51"，确定单片机类型，如图 1-14 所示。在随后的对话框中单击"OK"按钮，为项目添加 STARTUP.A51 文件。

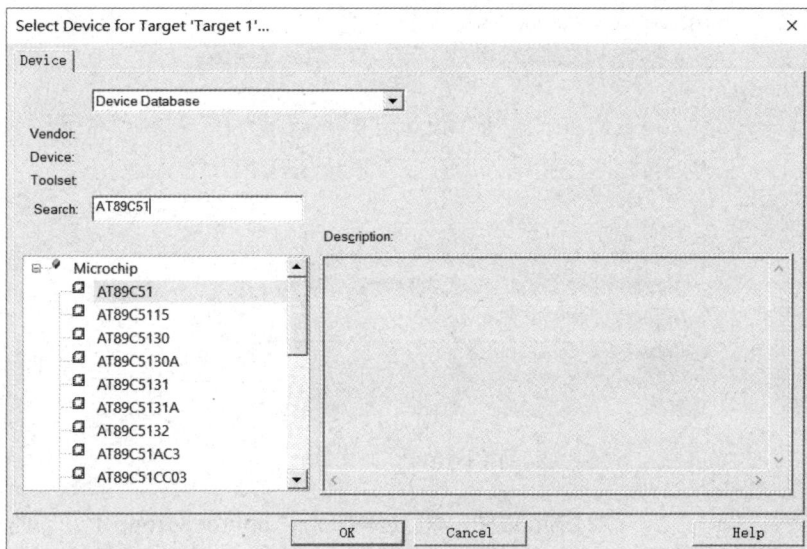

图 1-14　确定单片机类型

在 Keil μVision5 的菜单栏中执行"File"→"New"命令，新建文档，执行"File"→"Save"命令，保存文档，这时会弹出"Save As"对话框，如图 1-15 所示，在"文件名"输入框中为此文档命名，注意要填写扩展名".C"。

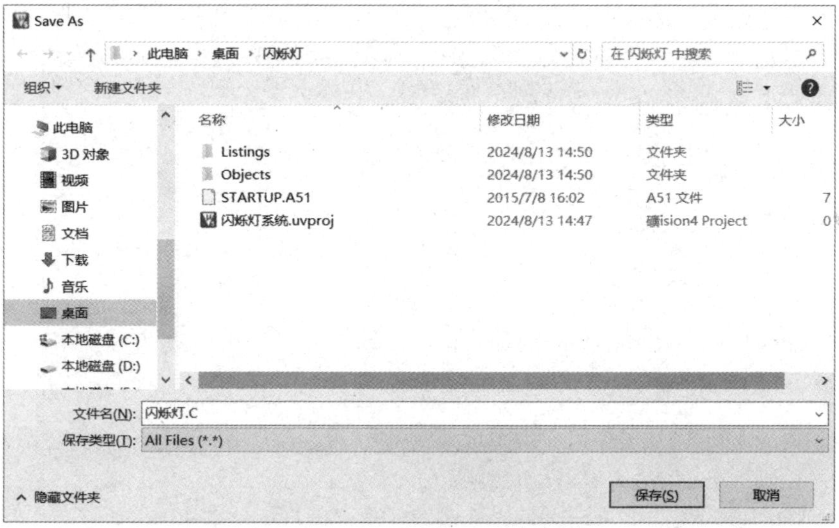

图 1-15　"Save As"对话框

单击"保存"按钮，这样在编写代码时，Keil 会自动识别语句的关键字，并以不同的颜色显示，以减少在输入代码时出现的语法错误。

当程序编写完后，必须再次保存。在 Keil μVision5 中的"Project"子窗口中，单击"Target 1"文件夹前的"+"按钮，展开。右击"Source Group 1"文件夹，在弹出的快捷菜单中选择"Add Files to Group'Source Group 1'"选项，弹出"Add Files to Group'Source Group'"对话框，找到刚才新建的"闪烁灯.C"文件，如图 1-16 所示。

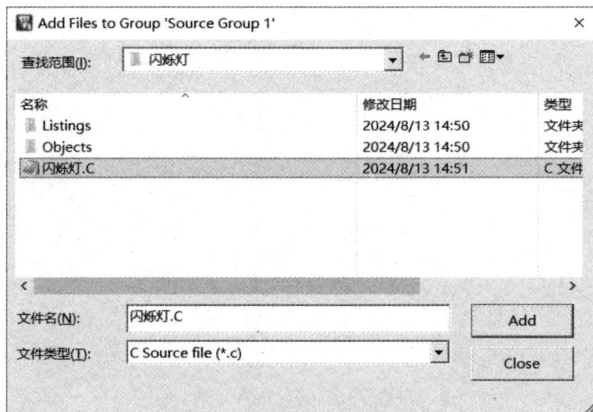

图 1-16　添加文件

图 1-17　"Project"子窗口

双击此文件，将其添加到"Source Group 1"文件夹中，此时的"Project"子窗口如图 1-17 所示。

右击"Project"子窗口中的"Target 1"文件夹，在弹出的快捷菜单中选择"Options for Target 'Target 1'"选项，这时会弹出"Options for Target 'Target 1'"对话框，在此对话框中选择"Output"选项卡，勾选"Create HEX File"复选框，如图 1-18 所示。

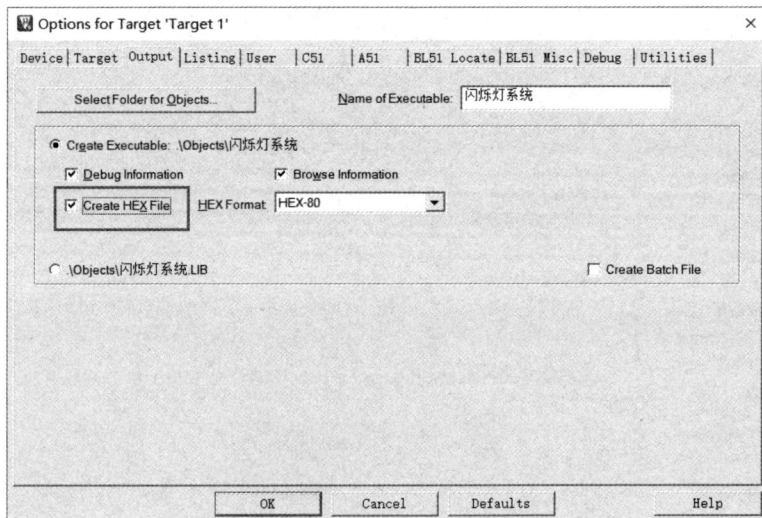

图 1-18　"Options for Target 'Target 1'"对话框

（3）参考程序。

```
/**********************************************************
程序名称：program1-1.C
程序功能：闪烁灯程序
**********************************************************/
#include<reg52.h>                              //加载头文件
```

闪烁灯参考程序

```
/**********************************************************************
数据类型定义
**********************************************************************/
#define uchar unsigned char                    //定义无符号字符型
/**********************************************************************
单片机引脚定义
**********************************************************************/
sbit LED=P0^0;                                  //定义 I/O 接口
/**********************************************************************
函数名称：延时函数
功能描述：延时 t ms（晶振频率为 12MHz）
入口参数：t
**********************************************************************/
void delay(uchar t)
{
    uchar i;
    while(t--)
    {
        i=123;
        while(i--);                             //实现 1ms 延时
    }
}
/**********************************************************************
主程序
**********************************************************************/
void main()
{
    while(1)                                    //重复循环
    {
        LED=0;                                  //点亮 LED
        delay(200);                             //延时 200ms
        LED=1;                                  //熄灭 LED
        delay(200);                             //延时 200ms
    }
}
```

闪烁灯仿真效果视频

（4）编译程序。

输入程序后，在 Keil μVision5 的菜单栏中执行"Project"→"Build Target"命令，编译程序。如果编译成功，则在 Keil μVision5 的"Build Output"子窗口中会显示图 1-19 所示的信息；如果编译不成功，则双击"Build Output"子窗口中的错误信息，在编辑窗口中指示错误的语句。

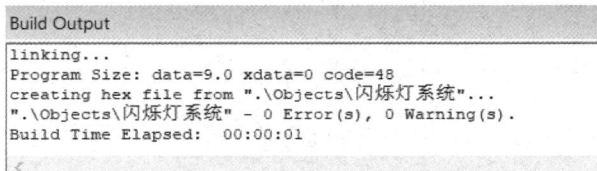

```
Build Output

linking...
Program Size: data=9.0 xdata=0 code=48
creating hex file from ".\Objects\闪烁灯系统"...
".\Objects\闪烁灯系统" - 0 Error(s), 0 Warning(s).
Build Time Elapsed:  00:00:01
<
```

图 1-19　编译程序

3. 系统仿真

Keil C51 编译成功后，会自动产生 HEX 文件，接着打开之前绘制的 Proteus 仿真图，双击 AT89C51，弹出图 1-20 所示的对话框，单击"Program File"中的文件夹按钮，在弹出的"Select File Name"对话框中，选择之前编译生成的 HEX 文件，单击"打开"按钮，返回"Edit Component"对话框，单击"OK"按钮，即可装入 HEX 文件。

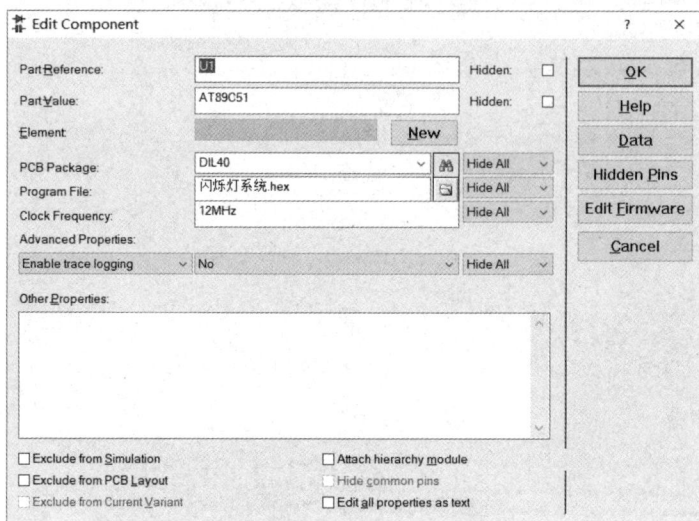

图 1-20 装入 HEX 文件

装入 HEX 文件后，单击 Proteus ISIS 软件界面左下角的运行按钮▶，即可观察发光二极管是否能够实现闪烁灯的功能，也可在 P0.0 口添加示波器，如图 1-21 所示，验证 P0.0 口的输出波形是否为周期 400ms 的方波（发光二极管亮 200ms、灭 200ms），示波器的波形图如图 1-22 所示。

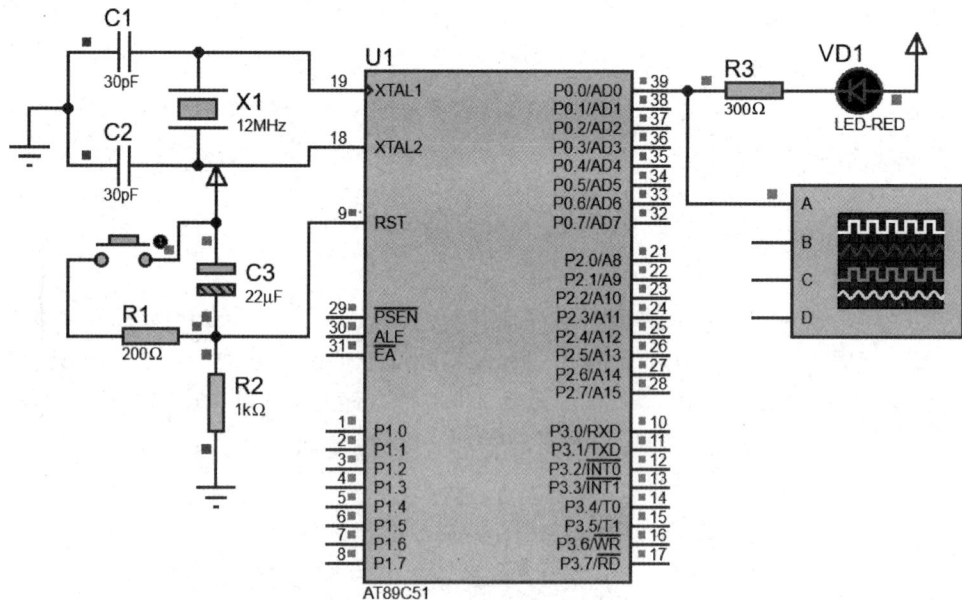

图 1-21 添加示波器

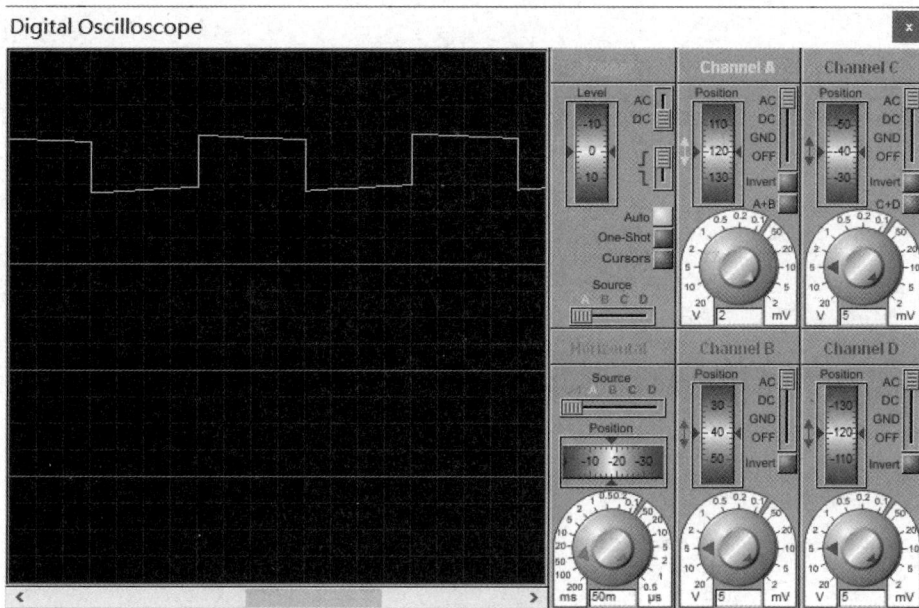

图 1-22 示波器的波形图

任务 1.2 简易信号灯的仿真设计

简易信号灯导学材料

学习目标

【知识目标】

（1）了解并掌握 C51 的基础知识。

（2）了解并掌握 intrins.h 头文件的定义及使用方法。

（3）了解并掌握单片机 I/O 接口的使用方法。

【技能目标】

（1）了解并掌握单片机仿真软件 Proteus 的使用方法。

（2）了解并掌握单片机编译软件 Keil C51 的使用方法。

（3）通过简易信号灯的仿真设计进一步掌握单片机项目的开发步骤。

【思政目标】

（1）培养学生树立正确的技能观，努力提高自己的职业技能。

（2）培养学生采用多种程序设计方法设计简易信号灯，从而让学生理解"条条大路通罗马"的哲学道理。

1.2.1 C51 的基础知识二

1. 算术运算符

在 C51 语言中，算术运算符主要有加、减、乘、除、取余等，如表 1-5 所示。

C51 运算符的使用（一）微课视频

表 1-5　算术运算符

符号	功能	范例	说明
+	加	a=x+y	将变量 x 与变量 y 的值相加后，放入变量 a
−	减	a=x-y	将变量 x 与变量 y 的值相减后，放入变量 a
*	乘	a=x*y	将变量 x 与变量 y 的值相乘后，放入变量 a
/	除	a=x/y	将变量 x 的值除以变量 y 的值，放入变量 a
%	取余	a=x%y	将变量 x 的值除以变量 y 的值，余数放入变量 a

算术运算符的优先级：先乘、除、取余，后加、减，括号最优先。即在算术运算符中，乘、除、取余的优先级是相同的，并高于加、减的优先级。在表达式中若出现括号，则括号的优先级最高。

2. 关系运算符

在 C51 语言中，关系运算符主要有等于、不等于、大于、小于、大于或等于、小于或等于等，如表 1-6 所示。

表 1-6　关系运算符

符　　号	功　　能	范　　例	说　　明
==	等于	x==y	比较变量 x 和变量 y 的值是否相等，相等则为 1，不相等则为 0
!=	不等于	x!=y	比较变量 x 和变量 y 的值是否相等，不相等则为 1，相等则为 0
>	大于	x>y	若变量 x 的值大于变量 y 的值，则为 1，否则为 0
<	小于	x<	若变量 x 的值小于变量 y 的值，则为 1，否则为 0
>=	大于或等于	x>=y	若变量 x 的值大于或等于变量 y 的值，则为 1，否则为 0
<=	小于或等于	x<=y	若变量 x 的值小于或等于变量 y 的值，则为 1，否则为 0

关系运算符的优先级有以下几种规定。

① 表 1-6 中的后四种关系运算符（>、<、>=、<=）的优先级相同，前两种关系运算符的优先级相同。

② 后四种关系运算符（>、<、>=、<=）的优先级高于前两种关系运算符的优先级。

③ 关系运算符的优先级低于算术运算符的优先级。

④ 关系运算符的优先级高于赋值运算符的优先级。

3. for 语句

在 C51 语言中，for 语句是循环语句中最为灵活也最为复杂的一种。它不仅可以用于循环次数已经确定的情况，还可以用于循环次数不确定但已给出循环条件的情况。它既可以包含一个索引计数变量，又可以包含任意一种表达式。其一般形式如下：

```
for(表达式 1; 表达式 2; 表达式 3)
{
    语句; //循环体
}
```

for 语句的执行过程如图 1-23 所示，具体如下。

C51for 语句的使用微课视频

① 计算“表达式 1”的值。

② 计算“表达式 2”的值（设为 x），若其值非 0，则转入步骤③；若其值为 0，则转入步骤⑤。

③ 执行循环体。

④ 计算“表达式 3”的值，然后转入步骤②。

⑤ 退出 for 循环，执行下面的语句。

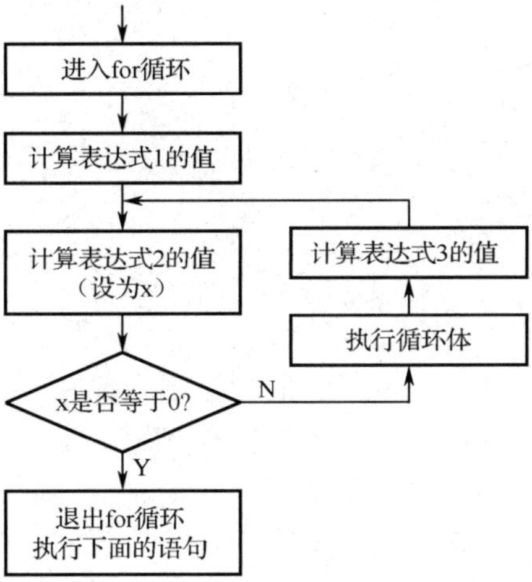

图 1-23　for 语句的执行过程

在使用 for 语句时应该注意以下两点。

① for 语句中的表达式可以部分省略或全部省略，但两个“;”不可省略。

例如：for(; ;)语句的 3 个表达式均省略，但因缺少条件判断，循环将会无限制地执行，而形成无限循环（通常称为死循环）。此时，等同于 while(1)语句。

② 所谓省略只是在 for 语句中的省略。实际上是把所需的表达式挪到循环体中或 for 语句之前。

例如，以下几种 for 语句的表达方式是等价的。

表达方式 1（正常情况）：

```
j=0;
for(i=0; i<100; i++)
      j++;
```

表达方式 2（省略表达式 1）：

```
j=0;
i=0;
for(; i<100; i++)
      j++;
```

表达方式 3（省略表达式 3）：

```
j=0;
for(i=0;i<100; )
{
    j++;
    i++;
}
```

表达方式 4（省略表达式 1 和表达式 3）：

```
j=0;
i=0;
for(; i<100; )
{
    j++;
    i++;
}
```

4. 三种循环语句的比较

while、do while、for 都是循环语句，它们在使用时有以下特点。

① 三种循环语句一般可以相互替代处理同一问题。

② do while 语句至少执行一次循环体，而其余两种循环语句则不然。

③ while 和 do while 语句多用于循环次数不可预知的情况，而 for 语句多用于循环次数可预知的情况。

5. 循环嵌套

在一个循环体内完整地包含另一个循环体，称为循环嵌套。前面介绍的三种循环都可以互相嵌套，循环嵌套可以有多层，但每层循环在逻辑上必须是完整的。在编写程序时，循环嵌套的书写要采用缩进形式，使程序层次分明。

在进行循环嵌套时，应注意以下几点。

① 内外循环的循环变量应该不同。

② 内外循环不应交叉。

③ 只能从循环体内转移到循环体外，反之不行。

6. intrins.h 头文件

在 C51 单片机编程中，intrins.h 头文件的函数使用起来，会让你像使用汇编一样简便。intrins.h 头文件主要包含表 1-7 所示的内部函数说明。

表 1-7　intrins.h 头文件的内部函数说明

内部函数	功能说明	内部函数	功能说明
crol	字符循环左移（单字节）	_lrol_	长整数循环左移（四字节）
cror	字符循环右移（单字节）	_lror_	长整数循环右移（四字节）
irol	整数循环左移（双字节）	_nop_	空操作（相当于 8051 NOP 指令）
iror	整数循环右移（双字节）	_testbit_	测试并清零位（相当于 8051 JBC 指令）

例如，已知变量 a=0x8c=10001100B，执行语句

```
a=_crol_(a,1);//循环左移一次
```

则最终 a=00011001B=0x19。

例如，已知变量 a=0xac=10101100B，执行语句

```
a=_cror_(a,3);//循环右移三次
```

则最终 a=10010101B=0x95。

1.2.2　简易信号灯的任务实施

【设计要求】

在单片机的 P0.0～P0.3 口分别接 4 个不同颜色的发光二极管，其中 P0.0 口接红色发光二极管、P0.1 口接黄色发光二极管、P0.2 口接绿色发光二极管、P0.3 口接蓝色发光二极管，实现简易信号灯的效果，要求首先红灯点亮，1s 后自动熄灭，同时黄灯点亮，1s 后自动熄灭，绿灯点亮，1s 后自动熄灭，蓝灯点亮，1s 后自动熄灭，循环往复。

【任务分析】

在本次任务中，单片机采用低电平驱动的方式驱动发光二极管点亮。单片机的 I/O 接口可以按位操作编程，也可以按字节操作编程。按照设计要求，可以得出表 1-8 所示的简易信号灯的状态表。

表 1-8　简易信号灯的状态表

序号	时间	信号灯状态	P0 口的值
1	第 1s	红灯点亮，其余灯熄灭	P0=11111110B=0xfe
2	第 2s	黄灯点亮，其余灯熄灭	P0=11111101B=0xfd
3	第 3s	绿灯点亮，其余灯熄灭	P0=11111011B=0xfb
4	第 4s	蓝灯点亮，其余灯熄灭	P0=11110111B=0xf7

通过表 1-8，可以发现 P0 口的值每隔 1s 循环左移一次，因此也可以采用_crol_函数实现简易信号灯的效果。

【实施步骤】

1．添加元器件

打开 Proteus 仿真软件，按照表 1-9 添加元器件。

表 1-9　简易信号灯的元器件清单

元器件名称	所属类	所属子类
AT89C51	Microprocessor ICs	8051 Family
CRYSTAL	Miscellaneous	—
CAP	Capacitors	Generic
CAP-ELEC	Capacitors	Generic
RES	Resistors	Generic
BUTTON	Switches & Relays	Switches
LED-RED	Optoelectronics	LEDs

续表

元器件名称	所属类	所属子类
LED-YELLOW	Optoelectronics	LEDs
LED-GREEN	Optoelectronics	LEDs
LED-BLUE	Optoelectronics	LEDs

2. 绘制仿真图

元器件全部添加后，在 Proteus ISIS 的原理图编辑窗口中按图 1-24 绘制仿简易信号灯仿真图。

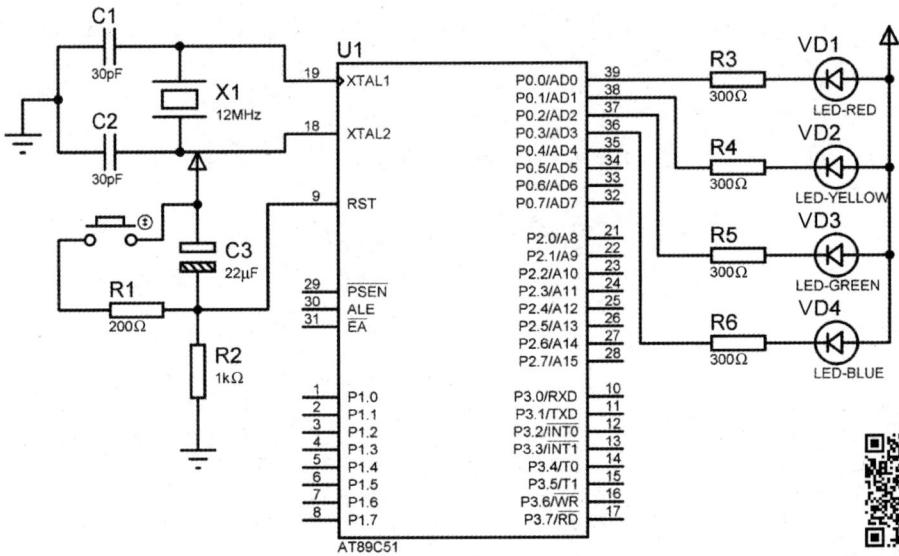

图 1-24　简易信号灯仿真图

简易信号灯程序
（第一种思路）

3. 编写程序

（1）第一种设计思路。

采用任务 1.1 的编程思路，逐个定义单片机的 I/O 接口，完成设计要求。程序包含主程序、延时函数、显示函数三部分。其中，显示函数的流程图如图 1-25 所示。

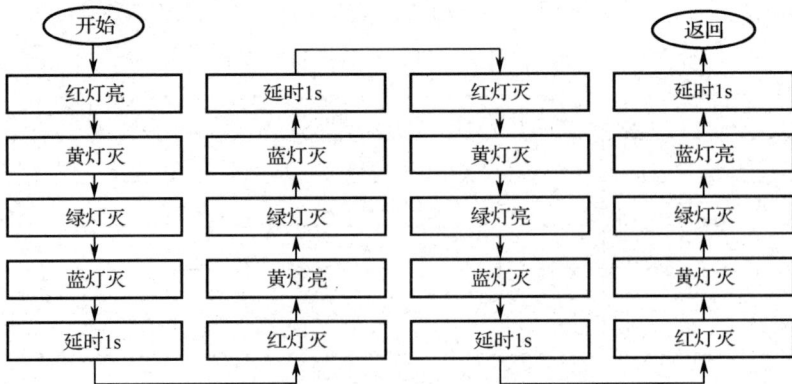

图 1-25　显示函数的流程图（第一种设计思路）

在 Keil μVision5 中编写程序，实现简易信号灯的效果，参考程序如下：

```c
/**********************************************************************
程序名称：program1-2.c
程序功能：简易信号灯程序（第一种设计思路）
**********************************************************************/
#include<reg52.h>                          //加载头文件
/**********************************************************************
数据类型定义
**********************************************************************/
#define uchar unsigned char                //定义无符号字符型
#define uint unsigned int                  //定义无符号整型
/**********************************************************************
单片机引脚定义
**********************************************************************/
sbit LED_RED=P0^0;                         //红灯
sbit LED_YELLOW=P0^1;                      //黄灯
sbit LED_GREEN=P0^2;                       //绿灯
sbit LED_BLUE=P0^3;                        //蓝灯
/**********************************************************************
函数名称：延时函数
功能描述：延时 t ms（晶振频率为 12MHz）
入口参数：t
**********************************************************************/
void delayms(uint t)
{
    uchar i;
    while(t--)
        for(i=96; i>0; i--);               //实现 1ms 延时
}
/**********************************************************************
函数名称：显示函数
功能描述：实现信号灯的显示
入口参数：无
**********************************************************************/
void display()
{
    LED_RED=0;                             //红灯亮
    LED_YELLOW=1;                          //黄灯灭
    LED_GREEN=1;                           //绿灯灭
    LED_BLUE=1;                            //蓝灯灭
    delayms(1000);                         //延时 1s
    LED_RED=1;                             //红灯灭
    LED_YELLOW=0;                          //黄灯亮
    LED_GREEN=1;                           //绿灯灭
```

```
        LED_BLUE=1;                        //蓝灯灭
        delayms(1000);                     //延时 1s
        LED_RED=1;                         //红灯灭
        LED_YELLOW=1;                      //黄灯灭
        LED_GREEN=0;                       //绿灯亮
        LED_BLUE=1;                        //蓝灯灭
        delayms(1000);                     //延时 1s
        LED_RED=1;                         //红灯灭
        LED_YELLOW=1;                      //黄灯灭
        LED_GREEN=1;                       //绿灯灭
        LED_BLUE=0;                        //蓝灯亮
        delayms(1000);                     //延时 1s
}
/*******************************************************************
主程序
*******************************************************************/
void main()
{
    while(1)                               //重复循环
        display();                         //调用显示函数
}
```

（2）第二种设计思路。

根据表 1-8，对单片机的 P0 口按字节操作编程，完成设计要求。程序包含主程序、延时函数、显示函数三部分。其中，显示函数的流程图如图 1-26 所示。

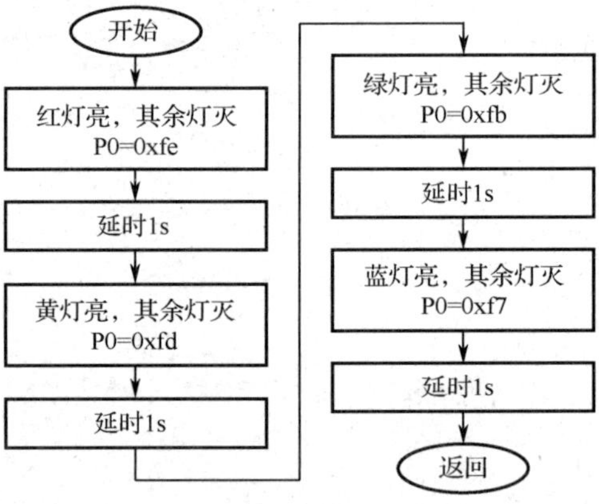

图 1-26 显示函数的流程图（第二种设计思路）

在 Keil μVision5 中编写程序，实现简易信号灯的效果，参考程序如下：

```
/*******************************************************************
程序名称：program1-3.c
程序功能：简易信号灯程序（第二种设计思路）
*******************************************************************/
```

```
#include<reg52.h>                          //加载头文件
/*****************************************************************
数据类型定义
*****************************************************************/
#define uchar unsigned char                //定义无符号字符型
#define uint unsigned int                  //定义无符号整型
/*****************************************************************
函数名称：延时函数
功能描述：延时 t ms（晶振频率为 12MHz）
入口参数：t
*****************************************************************/
void delayms(uint t)
{
    uchar i;
    while(t--)
        for(i=96; i>0; i--);               //实现 1ms 延时
}
/*****************************************************************
函数名称：显示函数
功能描述：实现信号灯的显示
入口参数：无
*****************************************************************/
void display()
{
    P0=0xfe;                               //红灯亮，其余灯灭
    delayms(1000);                         //延时 1s
    P0=0xfd;                               //黄灯亮，其余灯灭
    delayms(1000);                         //延时 1s
    P0=0xfb;                               //绿灯亮，其余灯灭
    delayms(1000);                         //延时 1s
    P0=0xf7;                               //蓝灯亮，其余灯灭
    delayms(1000);                         //延时 1s
}
/*****************************************************************
主程序
*****************************************************************/
void main()
{
    while(1)                               //重复循环
        display();                         //调用显示函数
}
```

简易信号灯程序
（第二种思路）

简易信号灯程序
（第三种思路）

（3）第三种设计思路。

根据表 1-8，对单片机的 P0 口按循环左移操作编程，加载 intrins.h 头文件，并使用 for 语句实现设计要求。程序包含主程序、延时函数、显示函数三部分。其中，显示函数的流程图如图 1-27 所示。

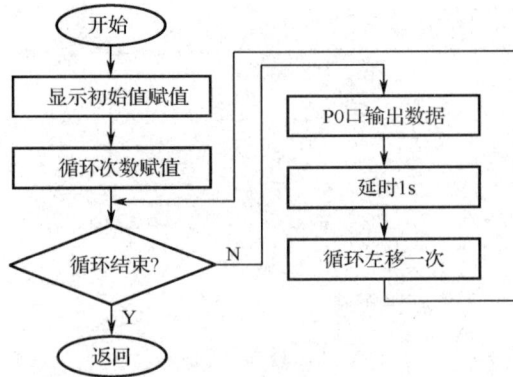

图 1-27　显示函数的流程图（第三种设计思路）

在 Keil μVision5 中编写程序，实现简易信号灯的效果，参考程序如下：

```
/****************************************************************************
程序名称：program1-4.c
简易信号灯程序（第三种设计思路）
****************************************************************************/
#include<reg52.h>                          //加载头文件
#include<intrins.h>                         //加载头文件
/****************************************************************************
数据类型定义
****************************************************************************/
#define uchar unsigned char                 //定义无符号字符型
#define uint unsigned int                   //定义无符号整型
/****************************************************************************
函数名称：延时函数
功能描述：延时 t ms（晶振频率为 12MHz）
入口参数：t
****************************************************************************/
void delayms(uint t)
{
    uchar i;
    while(t--)
        for(i=96; i>0; i--);                //实现 1ms 延时
}
/****************************************************************************
函数名称：显示函数
功能描述：实现信号灯的显示
入口参数：无
****************************************************************************/
void display()
{
    uchar i,j=0xfe;                         //显示初始值，红灯亮，其余灯灭
```

```
        for(i=4;i>0;i--)
        {
            P0=j;                              //点亮 LED
            Delayms(1000);                     //延时 1s
            j=_crol_(j,1);                     //循环左移点亮
        }
}
/*********************************************************************
主程序
*********************************************************************/
void main()
{
    while(1)                                   //重复循环
        display();                             //调用显示函数
}
```

4. 系统仿真

当 Keil C51 编译成功后，会自动产生 HEX 文件，接着打开之前绘制的 Proteus 仿真图，双击 AT89C51，弹出"Edit Component"对话框，单击"Program File"中的文件夹按钮，在弹出的"Select File Name"对话框中，选择之前编译生成的 HEX 文件，单击"打开"按钮，返回"Edit Component"对话框，单击"OK"按钮，即可装入 HEX 文件。

简易信号灯的
仿真效果视频

接着单击 Proteus ISIS 编辑界面左下角的运行按钮▷，即可观察 4 个发光二极管是否能够实现简易信号灯的功能，如图 1-28 所示。

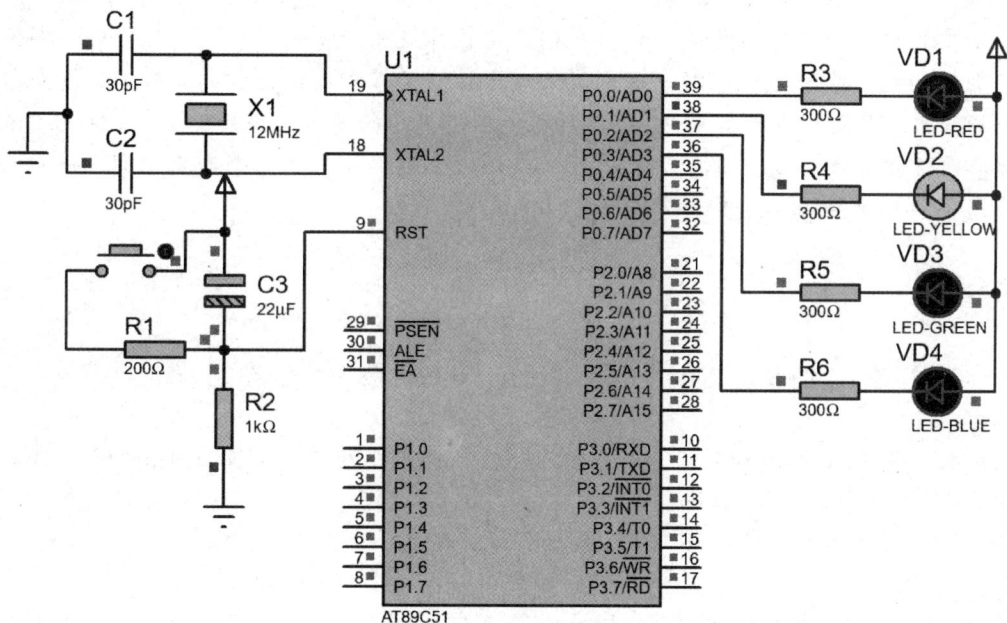

图 1-28　简易信号灯仿真图

素养小课堂

单片机的历史发展

单片机出现的历史并不长，它的产生与发展和微处理器的产生与发展大体上同步，经历了四个阶段。

第一阶段（1971—1974 年）：1971 年 11 月，美国 Intel 公司首先设计出集成度为 2000 只晶体管/片的 4 位微处理器——Intel 4004，并配有 RAM、ROM 和移位寄存器等芯片，构成了第一台 MCS-4 微型计算机。1972 年 4 月，Intel 公司又成功研制了处理能力较强的 8 位微处理器——Intel 8008。这些微处理器虽说还不是单片机，但从此拉开了研制单片机的序幕。

第二阶段（1974—1978 年）：初级单片机阶段。以 Intel 公司的 MCS-48 为代表，这个系列的单片机内集成有 8 位 CPU、I/O 接口、8 位定时器/计数器，寻址范围不大于 4KB，且无串行接口。

第三阶段（1978—1983 年）：高性能单片机阶段。在这一阶段推出的单片机普遍带有串行 I/O 接口，有多级中断处理系统、16 位定时器/计数器。单片机内 RAM、ROM 容量加大，且寻址范围可达 64KB，有的片内还带有 A/D 转换器接口。这类单片机有 Intel 公司的 MCS-51、Motorola 公司的 6801 和 Zilog 公司的 Z80 等。这类单片机的应用领域极其广泛，这个系列的各类产品仍然是目前国内外产品中的主流。其中，MCS-51 系列产品以其优良的性能价格比，成为我国广大科技人员的首选。

第四阶段（1983—现在）：8 位单片机巩固发展及 16 位单片机推出阶段。此阶段单片机的主要特征有一方面发展 16 位单片机及专用单片机；另一方面不断完善高档 8 位单片机，改善其结构，以满足不同的用户需要。

纵观单片机五十多年的发展历程，我们认为单片机今后将向多功能、高性能、高速度、低电压、低功耗、低价格、外围电路内装化及内存储器容量增加的方向发展。但其位数不一定会继续增加，尽管现在已经有 32 位单片机，但使用的并不多。今后的单片机将功能更强、集成度和可靠性更高、使用更方便。此外，专用化也是单片机的一个发展方向，针对某一用途的专用单片机将会越来越多。

课后任务

1. 修改任务 1.1 的设计要求，在单片机的 P1.0 口上接一个发光二极管，使发光二极管的闪烁周期为 1s。

2. 修改任务 1.2 的设计要求，4 组彩灯分别接 P2.0～P2.3 口。要求灯光效果如下：①红灯亮 1s，其余灯灭；②黄灯亮 1s，其余灯灭；③绿灯亮 1s，其余灯灭；④蓝灯亮 1s，其余灯灭；⑤红灯、黄灯亮 1s，其余灯灭；⑥绿灯、蓝灯亮 1s，其余灯灭；⑦红灯、绿灯亮 1 s，其余灯灭；⑧黄灯、蓝灯亮 1s，其余灯灭；⑨全亮 1s；⑩全灭 1s。重复循环。

3. 修改任务 1.2 的设计要求，4 组彩灯分别接 P0.0～P0.3 口。要求每组彩灯轮流闪烁 2次，每次的闪烁周期为 200ms，即亮 100ms、灭 100ms。

课后任务 1 仿真效果视频　　　　课后任务 2 仿真效果视频　　　　课后任务 3 仿真效果视频

知识拓展　单片机的分类

1. 按厂家分类

（1）美国的芯片公司。

① 英特尔公司（Intel），代表产品有 MCS-51 系列、MCS-96 系列等。

② 摩托罗拉公司（Motorola），代表产品有 6805 系列等。

③ 美国国家半导体公司（NS），代表产品有 NS8070 系列等。

④ 爱特梅尔公司（Atmel），代表产品有 AT89 系列、ATMEGA8 系列等。

⑤ 美国微芯科技公司（Microchip），代表产品有 PIC16 系列、PIC18 系列等。

⑥ 罗克韦尔自动化公司（Rockwell），代表产品有 PPS/1 系列。

⑦ 德州仪器公司（TI），代表产品有 MSP430 系列等。

（2）日本的芯片公司。

① 东芝公司（Toshiba），代表产品有 TMP 系列等。

② 日本电气公司（NEC），代表产品有 UCOM87 系列等。

③ 松下电器公司（Panasonic），代表产品有 MN 系列等。

④ 富士通公司（Fujitsu），代表产品有 MB88 系列。

⑤ 夏普公司（Sharp），代表产品有 SM×× 系列。

⑥ 日立公司（Hitachi），代表产品有 HD6301 系列等。

2. 按字长分类

字长是 CPU 的主要技术指标之一，指的是 CPU 一次能够并行处理的二进制位数。

（1）4 位单片机。

4 位单片机的控制能力较弱，CPU 一次只能处理 4 位二进制数。这类单片机常用于计数器、各种形态的智能单元，以及家用电器中的控制器。典型产品有 NEC 公司的 UPD75××、NS 公司的 COP400 系列、Panasonic 公司的 MN1400 系列、Rockwell 公司的 PPS/1 系列、Fujitsu 公司的 MB88 系列、Sharp 公司的 SM×× 系列、Toshiba 公司的 TMP47××× 系列等。

（2）8 位单片机。

8 位单片机的控制能力较强，品种最为齐全，其片内资源丰富和功能强大，主要在工业控制、智能仪表、家用电器和办公自动化系统中应用。典型产品有 Intel 公司的 MCS51 系列、Microchip 公司的 PIC16 系列和 PIC18 系列、Motorola 公司的 M68HC05 系列和 M68HC11 系列、Zilog 公司的 Z8 系列、Philips 公司的 P89C51XX 系列、Atmel 公司的 AT89 系列等。

（3）16 位单片机。

16 位单片机的控制能力较强，为 16 位 CPU，运算速度普遍高于 8 位单片机，有的单片机的寻址范围达到了 1MB。这类单片机主要用于过程控制、智能仪表、家用电器，以及计算机外围设备的控制器等。典型产品有 Intel 公司的 MCS-96 系列、Motorola 公司的 M68HC16

系列、NS 公司的 783××系列、TI 公司的 MSP430 系列等。

（4）32 位单片机。

32 位单片机的字长为 32 位，是单片机的顶级产品，具有极高的运算速度。典型产品有 Intel 公司的 MCS-80960 系列、Motorola 公司的 M68300 系列、Hitachi 公司的 SH 系列等。

习题

一、单选题

1．Intel 8051 单片机是（　　）单片机。

A．8 位　　　　　　　B．16 位　　　　　　　C．32 位　　　　　　　D．64 位

2．单片机应用系统包括（　　）两部分。

A．硬件系统和控制程序　　　　　　　　C．时钟电路和复位电路

B．运算器和控制器　　　　　　　　　　D．程序存储器和数据存储器

3．必须将控制程序下载到单片机的（　　）中，单片机才能工作。

A．RAM　　　　　　B．ROM　　　　　　C．控制器　　　　　　D．运算器

二、填空题

1．在 MCS-51 系列单片机中，内部没有 ROM 的单片机型号是＿＿＿＿＿＿＿＿；内部有 4KB 掩膜 ROM 的单片机型号是＿＿＿＿＿＿＿＿。

2．单片机的完整名称为＿＿＿＿＿＿＿＿，又称＿＿＿＿＿＿＿＿。

3．单片机由＿＿＿＿＿＿、＿＿＿＿＿＿、＿＿＿＿＿＿、＿＿＿＿＿＿、定时器/计数器和串行接口六部分组成。

4．在 LED 控制电路中，为了控制流过 LED 的电流大小，需要连接＿＿＿＿＿＿电阻。

5．DIP 是指＿＿＿＿＿＿，是常用的单片机封装形式。

三、简答题

1．什么是单片机？简述单片机的特点。

2．什么是单片机应用系统？

3．简述单片机应用系统的开发流程。

项目二　流水灯系统的仿真设计

任务 2.1　八路流水灯的仿真设计

学习目标

【知识目标】

（1）了解并掌握 C51 的基础知识——数组的使用方法。

（2）了解并掌握单片机 I/O 接口的使用方法。

【技能目标】

（1）了解并掌握单片机仿真软件 Proteus 的使用方法。

（2）了解并掌握单片机编译软件 Keil C51 的使用方法。

（3）了解并掌握单片机程序下载的方法。

（4）了解并掌握单片机最小系统的组成。

（5）通过八路流水灯的设计制作初步了解并掌握单片机项目的开发步骤。

【思政目标】

（1）通过电路设计、程序编写和仿真调试，培养学生具备高度的责任心和耐心。

（2）培养学生做事有规则，遵守社会秩序、企业秩序的好品德。

八路流水灯导学材料

2.1.1　C51 的基础知识三

1. 一维数组

C51 语言具有使用户能够定义一组有序数据的能力，这组有序的数据被称为数组。数组是一组具有固定数目和相同类型成分分量的有序集合，其成分分量的类型为该数组的基本类型。例如，整型变量的有序集合称为整型数组，字符型变量的有序集合称为字符型数组。这些整型或字符型变量是各自所属数组的成分变量，称为数组元素。

C51 数组的使用微课视频

构成一个数组的各元素必须是同一类型的变量，不允许在同一数组中出现不同类型的变量。

数组数据是用同一个名字的不同下标访问的，数组的下标放在方括号中，是从 0 开始的一组有序整数。例如，数组 a[i]，当 i=0, 1, 2, …, n 时，a[0]，a[1]，a[2]，…，a[n]分别是数组 a[i]的元素（或成员）。数组有一维、二维、三维和多维之分。常用的数组有一维数组、二维数组、字符数组等。

（1）一维数组的定义。

一维数组的定义格式如下：

类型说明符　　　数组名[整型表达式];

例如，unsigned char ch[10]定义了一个一维无符号字符型数组，该数组有 10 个元素，每个元素由不同的下标表示，分别为 ch[0]，ch[1]，ch[2]，…，ch[9]。注意：数组的第 1 个元素的下标是 0 而不是 1，即数组的第 1 个元素是 ch[0]，而数组的第 10 个元素是 ch[9]。

（2）一维数组的初始化。

数组中的元素，可以在程序运行期间，用循环和键盘输入语句进行复制。但这样做将耗费许多机器运行时间，对于大型数组，这种情况更加突出。对此，可以用数组初始化的方法加以解决。

所谓数组初始化，就是在定义数组的同时，给数组赋值。这项工作是在程序的编译过程中完成的。

对数组的初始化可用以下方法实现。

① 在定义数组时对数组中的全部元素赋值。

例如，int a[6]={0, 1, 2, 3, 4, 5}。

在上面进行的定义和初始化中，将 a 数组中的全部元素的初始值依次放在花括号内。这样，在初始化后，a[0]=0，a[1]=1，a[2]=2，a[3]=3，a[4]=4，a[5]=5。

② 只对数组中的部分元素初始化。

例如，char b[10]={0, 1, 2, 3, 4, 5}。

上面定义的 b 数组共有 10 个元素，但花括号内只有 6 个初始值，则数组中的前 6 个元素被赋值，而后 4 个元素的值为 0，即 b[6]=0，b[7]=0，b[8]=0，b[9]=0。

③ 在定义数组时，若不对数组中的全部元素赋予初始值，则数组中的全部元素被默认地赋值为 0。

例如，char a[10]={0}。则 a 数组中的 10 个元素 a[0]~a[9]全部被赋值为 0。

注意：常量数组可以存放在单片机的 ROM 中，也可以存放在 RAM 中，变量数组必须存放在 RAM 中。若常量数组存放在 ROM 中，则定义数组时需要加上"code"，如 char code a[5]。若不加"code"，则表示常量数组存放在 RAM 中。

2. 二维数组

（1）二维数组的定义。

二维数组的定义格式如下：

类型说明符　　　数组名[整型表达式] [整型表达式];

例如，unsigned char a[3][5]定义了一个 3 行 5 列共 15 个元素的二维无符号字符型数组。

二维数组的存取顺序：按行存取，先存取第一行元素的第 0 列、第 1 列、第 2 列……直到最后一列，然后存取第二行的第 0 列、第 1 列……直到最后一列。如此顺序下去，直到最后一行的最后一列。

（2）二维数组的初始化。

① 分行给二维数组中的全部元素赋予初始值。

例如：

```
int a[3][4]=
{
```

```
    {1,2,3,4},
    {5,6,7,8},
    {9,10,11,12}
};
```

这种赋值方法很直观，把第一行花括号内的数据赋值给第一行元素，第二行花括号内的数据赋值给第二行元素……即按行赋予初始值。

② 也可以将所有数据写在一个花括号中，按数组的排列顺序对各元素赋予初始值。

例如，int a[3][4]={1,2,3,4,5,6,7,8,9,10,11,12}。

③ 对数组中的部分元素赋予初始值。

例如，int a[3][4]={{1},{0,5},{3,0,7}}。

这种赋值等价于下面的数组赋值：

```
int a[3][4]=
{
    {1,0,0},
    {0,5,0},
    {3,0,7}
};
```

3. 字符数组

（1）字符数组的定义。

字符数组的定义与数组的定义类似。例如，char a[5]定义了一个有 5 个元素的一维字符数组。

（2）字符数组置初始值。

字符数组置初始值的最直接的方法是将各字符逐个赋给数组中的各个元素。例如：

```
char a[10]={'B', 'E', 'a', '8', 'f', 'g', 'H', 'K', 'M', 'n' }
```

定义了一个字符数组 a[]，其有 10 个元素，用单引号' '括起来的字符为字符的 ASCII 码，而不是字符串，如'a'表示 a 的 ASCII 码为 97。

4. 逻辑运算符

在 C51 语言中，逻辑运算符主要有与运算、或运算、反相运算等，如表 2-1 所示。

C51 运算符的使用（二）

表 2-1 逻辑运算符

符号	功能	范例	说明
&&	与运算	(x>y)&&(y>z)	若变量 x 的值大于变量 y 的值，且变量 y 的值大于变量 z 的值，则其结果为 1，否则为 0
||	或运算	(x>y)||(y>z)	若变量 x 的值大于变量 y 的值，或者变量 y 的值大于变量 z 的值，则其结果为 1，否则为 0
|	反相运算	|(x>y)	若变量 x 的值大于变量 y 的值，则其结果为 0，否则为 1

5. 布尔运算符

在 C51 语言中，布尔运算符主要有按位与、按位或、按位异或、位取反、位左移、位右移等，如表 2-2 所示。

表 2-2　布尔运算符

符号	功能	范例	说明
&	按位与	a=x&y	将变量 x 和变量 y 的每位进行与运算，结果放入变量 a
|	按位或	a=x|y	将变量 x 和变量 y 的每位进行或运算，结果放入变量 a
^	按位异或	a=x^y	将变量 x 和变量 y 的每位进行异或运算，结果放入变量 a
~	位取反	a=~x	将变量 x 的值进行取反运算，结果放入变量 a
<<	位左移	a=x<<n	将变量 x 的值左移 n 位，最右边补 0，结果放入变量 a
>>	位右移	a=x>>n	将变量 x 的值右移 n 位，最左边补 0，结果放入变量 a

6. 运算符的优先级

在 C51 语言中有许多运算符，它们之间存在优先级，如表 2-3 所示。

表 2-3　运算符的优先级

优先级	运算符或操作符号	说明
1	( )	小括号
2	~、!	取反、反相运算
3	++、--	自增、自减
4	*、/、%	乘、除、取余
5	+、-	加、减
6	<<、>>	位左移、位右移
7	<、>、<=、>=、==、!=	关系运算
8	&	按位与
9	^	按位异或
10	|	按位或
11	&&	与运算
12	||	或运算
13	=	赋值运算

2.1.2　八路流水灯的任务实施

【设计要求】

通过单片机设计一个八路流水灯电路，要求八路流水灯的点亮效果为从上到下点亮，并重复循环。

八路流水灯仿真电路的绘制微课视频

【任务分析】

通过设计要求，需要在 P2 口产生图 2-1 所示的时序图。注意：单片机采用低电平驱动发光二极管。

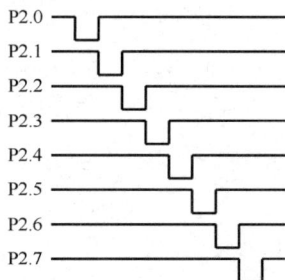

图 2-1　八路流水灯的时序图

【实施步骤】

1. 添加元器件

打开 Proteus 仿真软件，按照表 2-4 添加元器件。

表 2-4　八路流水灯的元器件清单

元器件名称	所属类	所属子类
AT89C51	Microprocessor ICs	8051 Family
CRYSTAL	Miscellaneous	—
CAP	Capacitors	Generic
CAP-ELEC	Capacitors	Generic
RES	Resistors	Generic
BUTTON	Switches & Relays	Switches
LED-RED	Optoelectronics	LEDs

2. 绘制仿真图

元器件全部添加后，在 Proteus ISIS 的原理图编辑窗口中按图 2-2 绘制八路流水灯仿真图。

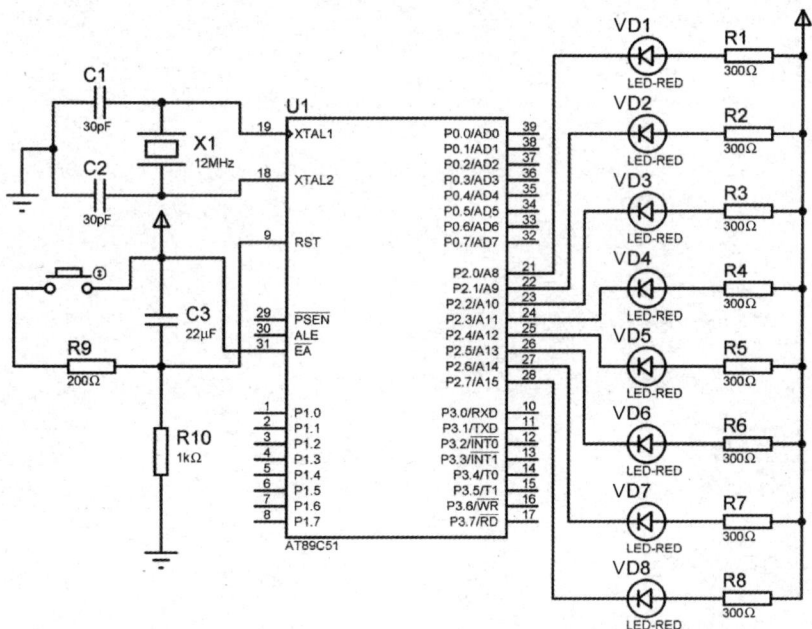

图 2-2　八路流水灯仿真图

3. 编写程序

（1）第一种设计思路。

采用对 P2 口循环左移的方式，完成设计要求。程序包含主程序、延时函数两部分。其中，主程序的流程图如图 2-3 所示。

八路流水灯程序
（第一种设计思路）

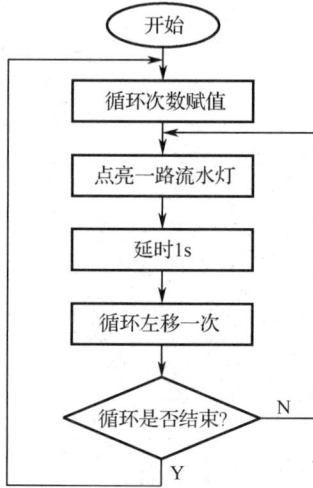

图 2-3　主程序的流程图（第一种设计思路）

在 Keil μVision5 中编写程序，实现八路流水灯的效果，参考程序如下：

```
/****************************************************************
程序名称：program2-1.c
程序功能：八路流水灯程序（第一种设计思路）
****************************************************************/
#include<reg52.h>                          //加载头文件
#include<intrins.h>
/****************************************************************
数据类型定义
****************************************************************/
#define uchar unsigned char                //定义无符号字符型
/****************************************************************
函数名称：延时函数
功能描述：延时 t ms（晶振频率为 12MHz）
入口参数：t
****************************************************************/
void delayms(uint t)
{
    uchar i;
    while(t--)
        for(i=96;i>0;i--);                 //实现 1ms 延时
}
/****************************************************************
```

主程序

```
*******************************************************************/
void main()                                    //主函数
{
    uchar i,j=0xfe;                            //点亮第一个 LED
    while(1)
    {
        for(i=8;i>0;i--)
        {
            P2=j;                              //P2 口输出点亮数据
            delayms(200);                      //延时 200ms
            j=_crol_(j,1);                     //左移
        }
    }
}
```

（2）第二种设计思路。

采用数组的方式对 P2 口赋值，实现八路流水灯的效果，如表 2-5 所示。

八路流水灯程序
（第二种设计思路）

表 2-5　八路流水灯状态取值

序号	LED8 P2.7	LED7 P2.6	LED6 P2.5	LED5 P2.4	LED4 P2.3	LED3 P2.2	LED2 P2.1	LED1 P2.0	P2 口取值
1	灭	灭	灭	灭	灭	灭	灭	亮	P2=11111110B=0xfe
2	灭	灭	灭	灭	灭	灭	亮	灭	P2=11111101B=0xfd
3	灭	灭	灭	灭	灭	亮	灭	灭	P2=11111011B=0xfb
4	灭	灭	灭	灭	亮	灭	灭	灭	P2=11110111B=0xf7
5	灭	灭	灭	亮	灭	灭	灭	灭	P2=11101111B=0xef
6	灭	灭	亮	灭	灭	灭	灭	灭	P2=11011111B=0xdf
7	灭	亮	灭	灭	灭	灭	灭	灭	P2=10111111B=0xbf
8	亮	灭	灭	灭	灭	灭	灭	灭	P2=01111111B=0x7f

在 Keil μVision5 中编写程序，实现八路流水灯的效果，参考程序如下：

```
/********************************************************************
程序名称：program2-2.c
程序功能：八路流水灯程序（第二种设计思路）
********************************************************************/
#include<reg52.h>                             //加载头文件
/********************************************************************
数据类型定义
********************************************************************/
#define uchar unsigned char                   //定义无符号字符型
/********************************************************************
```

数组定义，将数组元素存放在单片机 ROM 中
***/

```c
uchar code DISPLAY_CODE[8]=
{
    0xfe,0xfd,0xfb, 0xf7, 0xef, 0xdf, 0xbf, 0x7f
};
/*******************************************************************
函数名称：延时函数
功能描述：延时 t ms（晶振频率为 12MHz）
入口参数：t
*******************************************************************/
void delayms(uint t)
{
    uchar i;
    while(t--)
        for(i=96;i>0;i--);                          //实现 1ms 延时
}
/*******************************************************************
主程序
*******************************************************************/
void main()                                         //主函数
{
    uchar i;
    while(1)
    {
        for(i=0;i<8;i++)
        {
            P2=DISPLAY_CODE[i];                     //P2 口输出点亮数据
            delayms(200);                           //延时 200ms
        }
    }
}
```

4. 系统仿真

当 Keil C51 编译成功后，会自动产生 HEX 文件，接着打开之前绘制的 Proteus 仿真图，双击 AT89C51，弹出"Edit Component"对话框，单击"Program File"中的文件夹按钮，在弹出的"Select File Name"对话框中，选择之前编译生成的 HEX 文件，单击"打开"按钮，返回"Edit Component"对话框，单击"OK"按钮，即可装入 HEX 文件。

八路流水灯仿真效果视频

接着单击 Proteus ISIS 编辑界面左下角的运行按钮 ▶ ，即可观察是否能够实现八路流水灯从上到下点亮的效果，如图 2-4 所示。

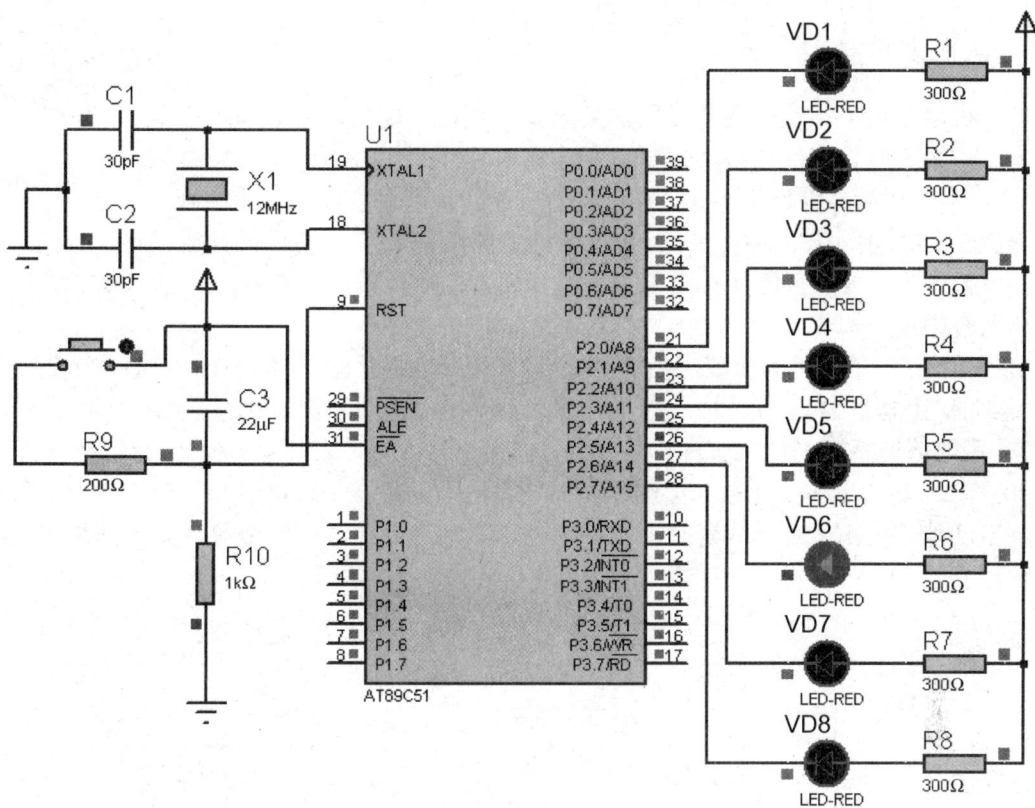

图 2-4 八路流水灯仿真效果图

任务 2.2 心形流水灯的仿真设计

学习目标

【知识目标】

（1）了解并掌握 C51 的基础知识。

（2）了解并掌握定时器的使用方法。

（3）了解并掌握单片机 I/O 接口的使用方法。

【技能目标】

（1）了解并掌握单片机仿真软件 Proteus 的使用方法。

（2）了解并掌握单片机编译软件 Keil C51 的使用方法。

（3）掌握定时器的使用方法。

（4）通过心形流水灯的仿真设计进一步掌握单片机项目的开发步骤。

【思政目标】

通过心形流水灯的仿真设计，培养学生在完成作品的过程中，不断追求完美，注重细节，提升作品的质量和性能。

心形流水灯导学材料

2.2.1 认识定时器/计数器

单片机应用技术往往需要定时检测某个参数或按一定的时间间隔进行某种控制。这种定时作用的获得固然可以利用延时程序来实现，但这样做是以降低 CPU 的工作效率为代价的。如果能通过一个可编程的实时时钟来实现，就不会影响 CPU 的工作效率。另外，还有一些控制是按对某种事件的计数结果进行的。因此在单片机系统中，常用到定时器/计数器。几乎所有的单片机内部都有定时器/计数器，这无疑可简化系统的设计。

MCS-51 系列单片机的典型产品 8051 有 2 个 16 位定时器/计数器 T0、T1；8052 有 3 个 16 位定时器/计数器 T0、T1、T2。它们都可以用作定时器或外部事件计数器。

1. 定时器的结构

8051 单片机内部有 2 个 16 位定时器/计数器，即定时器 T0 和定时器 T1。它们都具有定时和计数功能，可用于定时或延时控制，对外部事件进行检测、计数等。其内部结构框图如图 2-5 所示。

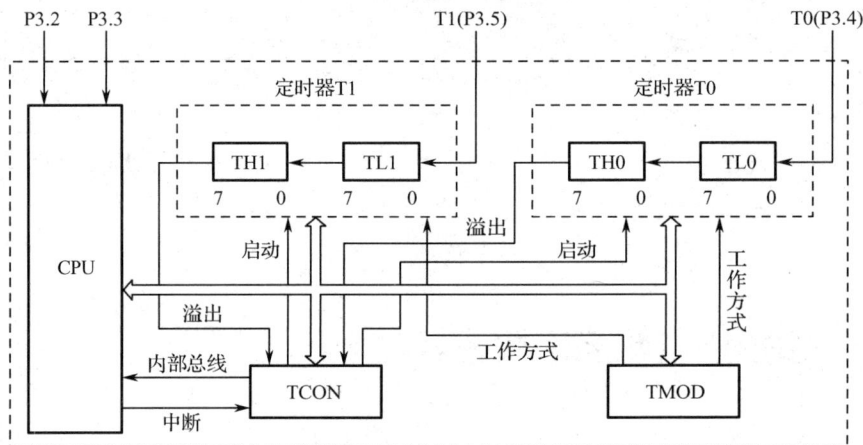

图 2-5　定时器/计数器的内部结构框图

定时器 T0 由 2 个特殊功能寄存器 TH0 和 TL0 构成，定时器 T1 由 TH1 和 TL1 构成。TH0 和 TL0 用来存放定时器 T0 的计数初始值，TH0 为高位；TH1 和 TL1 用来存放定时器 T1 的计数初始值，TH1 为高位。定时器方式选择寄存器（TMOD）用于设置定时器的工作方式，定时器控制寄存器（TCON）用于启动和停止计数，并控制定时器的状态。每个定时器的内部结构实质上都是一个可程控（程序控制）的加法计数器，由编程来设置它是工作在定时状态还是计数状态。

当设置了定时器的工作方式并启动定时器后，定时器就按设置的工作方式独立工作，不再占用 CPU 运行程序的操作，只有在计数器计满溢出时，才可能中断 CPU 当前的操作。用户可以重新设置定时器的工作方式，以改变定时器的状态。由此可见，定时器是单片机中工作效率高且应用灵活的部件。

2. TMOD

TMOD 的地址为 0x89，TMOD 不能支持位寻址，只能用字节指令设置定时器的工作方式，复位时，TMOD 所有位均为 0。TMOD 的格式如下。

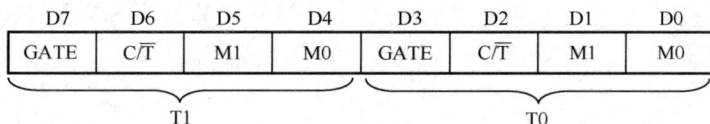

D7	D6	D5	D4	D3	D2	D1	D0
GATE	C/$\overline{T}$	M1	M0	GATE	C/$\overline{T}$	M1	M0

T1（D7~D4） T0（D3~D0）

TMOD 的低 4 位为 T0 的方式字段，高 4 位为 T1 的方式字段，它们的含义是完全相同的。

（1）门控位 GATE。

① 当 GATE=0 时，允许软件控制位 TR0 或 TR1 启动定时器。

② 当 GATE=1 时，允许外中断引脚电平启动定时器，即 $\overline{INT0}$（P3.2）和 $\overline{INT1}$（P3.3）引脚分别控制 T0 和 T1 的运行。

（2）功能选择位 C/$\overline{T}$。

① 当 C/$\overline{T}$=0 时，定时器/计数器的功能为定时。

② 当 C/$\overline{T}$=1 时，定时器/计数器的功能为计数。

（3）工作方式选择位 M1 和 M0。

M1 和 M0 的对应关系如表 2-6 所示。

表 2-6 M1 和 M0 的对应关系

M1	M0	工作方式	功能说明
0	0	方式 0	13 位定时器/计数器
0	1	方式 1	16 位定时器/计数器
1	0	方式 2	自动再装入 8 位定时器/计数器
1	1	方式 3	T0：分成 2 个 8 位计数器；T1：停止计数

例 1：（1）若 T0 采用方式 2 定时功能，则 TMOD 为多少？

（2）若 T1 采用方式 1 计数功能，则 TMOD 为多少？

解：（1）由于 T0 工作，所以 TMOD 的高 4 位均为 0，而 T0 的 GATE=0，C/$\overline{T}$=0，M1=1，M0=0。

所以，TMOD=00000010B=0x02。

（2）由于 T1 工作，所以 TMOD 的低 4 位均为 0，而 T1 的 GATE=0，C/$\overline{T}$=1，M1=0，M0=1。

所以，TMOD=01010000B=0x50。

3. TCON

TCON 的高 4 位用于存放定时器的运行控制位和溢出标志位，低 4 位用于存放外部中断的触发方式控制位和锁存外部中断请求源。TCON 的字节地址为 88H，支持位寻址。当单片机复位时，TCON 的所有位均为 0。下面主要介绍高 4 位，其各位的定义如表 2-7 所示，低 4 位将放在后面介绍。

表 2-7 TCON 高 4 位中各位的定义

位地址	8FH	8EH	8DH	8CH
位标志	TF1	TF1	TF0	TR0

（1）T1 溢出标志位 TF1。

当 T1 溢出时，硬件自动使 TF1 置 1，此时可以通过软件查询方式查看 TF1 的值，若 TF1=0，则表示定时时间还未达到；若 TF1=1，则表示定时时间已经达到。

（2）T1 启动标志位 TR1。

可由软件将 TR1 置 1 或清 0 来启动或关闭 T1。TR1=1 表示 T1 开始计数；TR1=0 表示 T1 停止计数。

（3）T0 溢出标志位 TF0，其功能及操作情况同 TF1。

（4）T0 启动标志位 TR0，其功能及操作情况同 TR1。

2.2.2 定时器/计数器的工作方式

MCS-51 系列单片机的 T0 有 4 种工作方式，即方式 0、方式 1、方式 2 和方式 3；而 T1 有 3 种工作方式，即方式 0、方式 1 和方式 2。

1. 方式 0

方式 0 为 13 位定时器/计数器，计数寄存器由 TH0（或 TH1）的全部 8 位和 TL0（或 TL1）的低 5 位构成，TL0（或 TL1）的高 3 位不用。当 TL0（或 TL1）的低 5 位计数溢出时向 TH0（或 TH1）进位，而 TH0（或 TH1）计数溢出时向 TF0（或 TF1）进位，并请求中断。因此，可通过查询 TF0（或 TF1）是否置位或考察中断是否发生（通过 CPU 响应）来判断 T0（或 T1）的操作是否完成。

（1）电路逻辑结构。

T0 的方式 0 的逻辑结构如图 2-6 所示。

在图 2-6 中，当 C/$\overline{\text{T}}$=0 时，多路开关接到振荡器的 12 分频输出，T0 对机器周期计数。

当 C/$\overline{\text{T}}$=1 时，多路开关与引脚 T0（P3.4）接通，T0 对来自外部引脚 T0（P3.4）的输入脉冲计数。当外部信号发生负跳变时，计数器加 1。

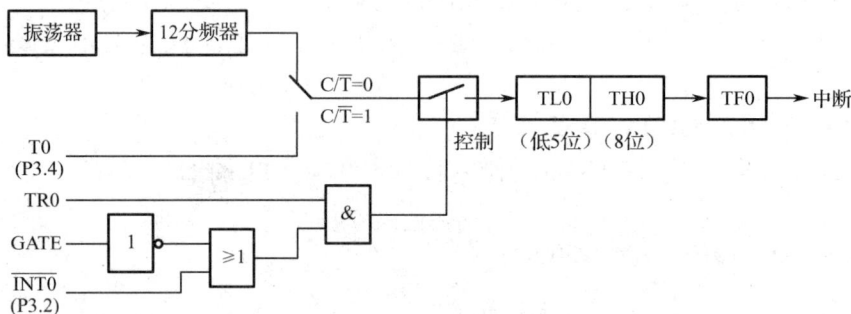

图 2-6 T0 的方式 0 的逻辑结构

GATE 控制定时器 T0 的运行条件：T0 取决于 TR0 的控制，还是取决于 TR0 和 $\overline{\text{INT0}}$ 的控制。

① 当 GATE=0 时，或门输出恒为 1，使外部中断输入引脚 $\overline{\text{INT0}}$ 信号失效，同时打开与门，由 TR0 控制定时器 T0 的开启和关断。若 TR0=1，则接通控制开关，启动定时器 T0 工作，计数器开始计数。若 TR0=0，则断开控制开关，计数器停止计数。

② 当 GATE=1 时，与门输出由 $\overline{\text{INT0}}$ 的输入电平和 TR0 的状态来确定。若 TR0=1，则打开与门，外部信号电平通过 $\overline{\text{INT0}}$ 引脚直接开启或关断定时器 T0。当 $\overline{\text{INT0}}$ 为高电平时，计数器开始计数，否则停止计数。这种工作方式可用来测量外部信号的脉冲宽度等。

以上说明同样适用于 T1。

（2）定时计算公式。

若 T0 计数初始值为 x，系统时钟频率为 f_{osc}，则其定时时间 t 为

$$t = (2^{13} - x) \times \frac{12}{f_{osc}}$$

注意：在给计数寄存器 TH0、TL0（或 TH1、TL1）赋初始值时，应将计算得到的计数初始值转换为二进制数，然后按其格式将低 5 位置入 TL0（或 TL1）的低 5 位，TL0（或 TL1）的高 3 位都可设为 0，而计数初始值的高 8 位则置入 TH0（或 TH1）。

例 2：现用 T1 做定时器，已知 f_{osc}=6MHz，采用方式 0，求最长的定时时间，以及相应的 T1 计数初始值分别为多少？

解：当 TH1=0x00，TL1=0x00 时，定时时间最长。

$$t = (2^{13} - x) \times \frac{12}{f_{osc}}$$

$$\Rightarrow t = (2^{13} - 0) \times \frac{12}{6 \times 10^6} = 16384 \text{（μs）}$$

2. 方式 1

T0 的方式 1 的逻辑结构如图 2-7 所示。

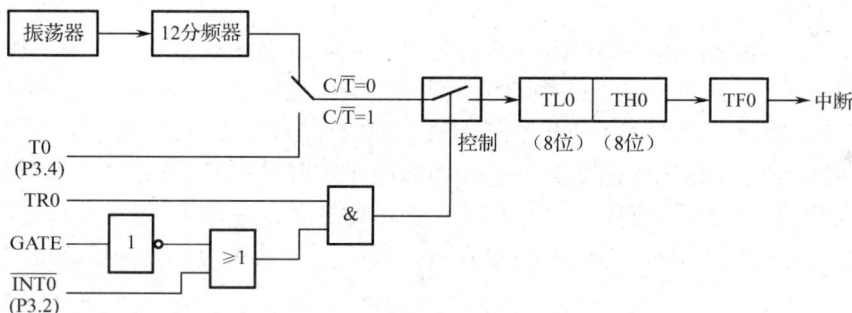

图 2-7　T0 的方式 1 的逻辑结构

方式 1 和方式 0 的差别仅在于计数器的位数不同。方式 1 为 16 位计数器，即 TH0 的 8 位和 TL0 的 8 位。

当 T0 作为定时器使用时，若其计数初始值为 x，系统时钟频率为 f_{osc}，则定时时间 t 为

$$t = (2^{16} - x) \times \frac{12}{f_{osc}}$$

例 3：现用 T0 做定时器，已知 f_{osc}=6MHz，采用方式 1，求最长的定时时间，以及相应的 T0 计数初始值分别为多少？

解：当 TH0=0，TL0=0 时，定时时间最长。

$$t = (2^{16} - x) \times \frac{12}{f_{osc}}$$

$$\Rightarrow t = (2^{16} - 0) \times \frac{12}{6 \times 10^6} = 131072 \text{（μs）}$$

3. 方式 2

T0 的方式 2 的逻辑结构如图 2-8 所示，它由作为 8 位计数器的 TL0 和作为重置初始值的缓冲器的 TH0 构成。T1 的方式 2 的逻辑结构与 T0 的类似。

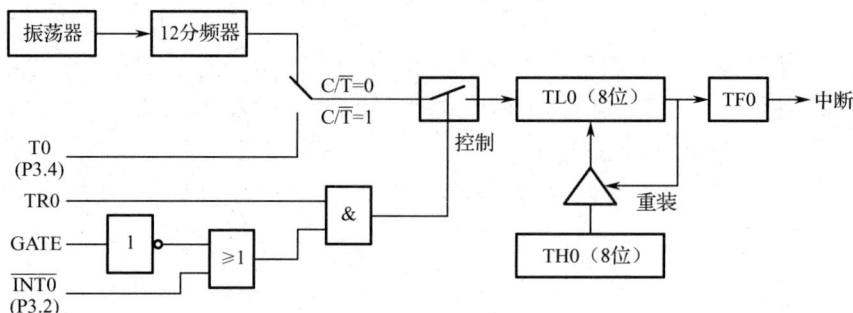

图 2-8　T0 的方式 2 的逻辑结构

若方式 0 和方式 1 用于循环重复定时或计数，则在每次计数满、溢出后，计数器清 0，要进行新一轮的计数就得重新装入计数初始值。这样不仅编程麻烦，而且影响定时时间的精度。而方式 2 具有自动装入初始值的功能，也就避免了上述缺点。因此它特别适合用作较精确的脉冲信号发生器。

在方式 2 中，16 位计数器被拆分成两部分：TL0 用作位计数器；TH0 用来保存计数初始值。在程序初始化时，由软件赋予同样的初始值。在操作过程中，一旦计数溢出，便置位 TF0，并将 TH0 中的初始值再装入 TL0，从而进入新一轮的计数，如此循环重复。

方式 2 可以避免在程序中因重新装入初始值而对定时精度产生的影响，适用于需要产生相当精度的定时时间的应用场合，常用于串行接口波特率发生器。

当 T0 用作定时器使用时，若 TH0=TL0= x，系统时钟频率为 f_{osc}，则其定时时间 t 为

$$t = (2^8 - x) \times \frac{12}{f_{osc}}$$

例 4：现用 T0 做定时器，已知 f_{osc}=6MHz，采用方式 2，求最长的定时时间，以及相应的 T0 的初始值分别为多少？

解：当 TH0=0，TL0=0 时，定时时间最长。

$$t = (2^8 - x) \times \frac{12}{f_{osc}}$$

$$\Rightarrow t = (2^8 - 0) \times \frac{12}{6 \times 10^6} = 512 \text{（μs）}$$

4. 方式 3

方式 3 的作用比较特殊，只适用于定时器 T0。如果企图将定时器 T1 置为方式 3，则它将停止计数，其效果与置 TR1=0 相同，即关闭定时器 T1。

当 T0 工作在方式 3 时，它被拆分成两个独立的 8 位计数器 TL0 和 TH0，其逻辑结构如图 2-9 所示。

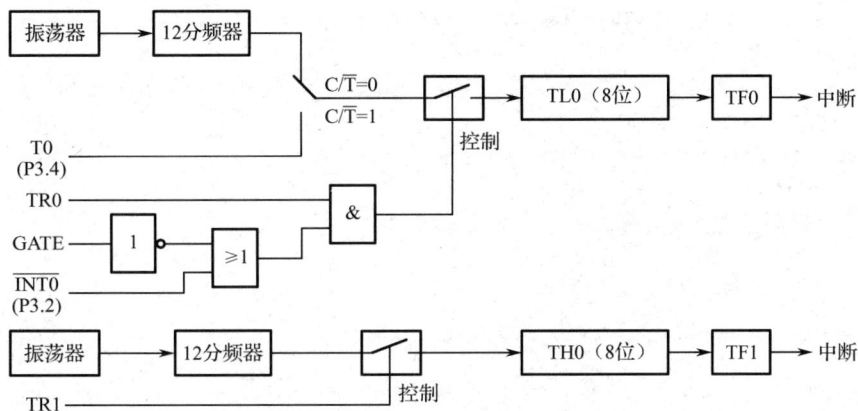

图 2-9　T0 的方式 3 的逻辑结构

在图 2-9 中，上方的 8 位计数器 TL0 使用原定时器 T0 的 $C/\overline{T}$、GATE、TR0 和 $\overline{INT0}$，它既可以定时（$C/\overline{T}$=0），又可以计数（$C/\overline{T}$=1）。而下方的 8 位计数器 TH0 占用了原定时器 T1 的 TR1 和 TF1，同时占用了 T1 中断源。它被固定为一个 8 位定时器，其启动和关闭仅受 TR1 的控制。当 TR1=1 时，控制开关接通，TH0 对 12 分频的时钟信号计数；当 TR1=0 时，控制开关断开，TH0 停止计数。由此可见，在方式 3 下，TH0 只能用于简单的内部定时，不能对外部脉冲进行计数，是定时器 T0 附加的一个 8 位定时器。

当定时器 T0 工作在方式 3 时，定时器 T1 仍可设置为方式 0、方式 1 或方式 2。但由于 TR1、TF1 及 T1 的中断源已被定时器 T0 占用，此时定时器 T1 仅由 $C/\overline{T}$ 切换定时或计数功能，当计数器计数满溢出时，只能将输出送往串行接口。在这种情况下，定时器 T1 一般用作串行接口波特率发生器或用于不需要中断的场合。

【课堂实训】编程实现单片机 P1.0 口输出周期为 2ms 的方波，要求采用定时器 T0 的方式 1 实现，已知 f_{osc}=12MHz。

分析：由于题目要求实现周期为 2ms 的方波，因此定时时间 t=1ms，采用定时器 T0 的方式 1 实现，因此 TMOD=0x01。

设定时器 T0 的初始值为 x，则

$$t=(2^{16}-x)\times\frac{12}{f_{osc}}$$

$$\Rightarrow 1\times10^{-3}=(2^{16}-x)\times\frac{12}{12\times10^{6}}$$

$$\Rightarrow x=64536=0xfc18$$

课堂实训
仿真效果视频

所以，TH0=0xfc，TL0=0x18。

参考程序如下：

```
/************************************************************
程序名称：program2-3.c
程序功能：P1.0 口输出周期为 2ms 的方波程序
```

```
*********************************************************************/
#include<reg52.h>                              //加载头文件
#define uchar unsigned char                    //定义无符号字符型
/*********************************************************************
引脚定义
*********************************************************************/
sbit OUTPUT=P1^0;                              //单片机输出引脚定义
/*********************************************************************
函数名称：延时函数
功能描述：T0 定时 1ms 函数（晶振频率为 12MHz）
入口参数：i，j
*********************************************************************/
void delay_t0(uchar i,j)                       //T0 定时 1ms 函数
{
    TH0=i;                                     //T0 定时 1ms 的初始值
    TL0=j;
    TR0=1;                                     //开启定时器 T0
    while(TF0==0);                             //判断 T0 是否溢出，若未溢出则等待
    TF0=0;                                     //若 T0 溢出，则清除 TF0
    TR0=0;                                     //关闭定时器 T0
}
/*********************************************************************
主函数
*********************************************************************/
void main()
{
    TMOD=0x01;                                 //T0 定时方式 1
    while(1)
    {
        delay_t0(0xfc,0x18);                   //T0 定时 1ms 函数
        OUTPUT=~OUTPUT;                        //波形翻转
    }
}
```

2.2.3 心形流水灯的任务实施

【设计要求】

通过单片机的 32 个 I/O 接口连接 32 个发光二极管，并将发光二极管布置成心形，实现心形流水灯的显示效果。

【任务分析】

在本次任务中，单片机采用低电平的方式驱动发光二极管点亮。单片机的 I/O 接口可以按位操作编程，也可以按字节方式操作编程。

【实施步骤】

1. 添加元器件

打开 Proteus 仿真软件，按照表 2-8 添加元器件。注意：用 Proteus 仿真软件绘制单片机仿真图时，可以省略振荡电路和复位电路。

表 2-8 心形流水灯的元器件清单

元器件名称	所属类	所属子类
AT89C51	Microprocessor ICs	8051 Family
RES	Resistors	Generic
CRYSTAL	Miscellaneous	—
CAP	Capacitors	Generic
CAP-ELEC	Capacitors	Generic
LED-GREEN	Optoelectronics	LEDs

2. 绘制仿真图

元器件全部添加后，在 Proteus ISIS 的原理图编辑窗口中按图 2-10 绘制心形流水灯仿真图。

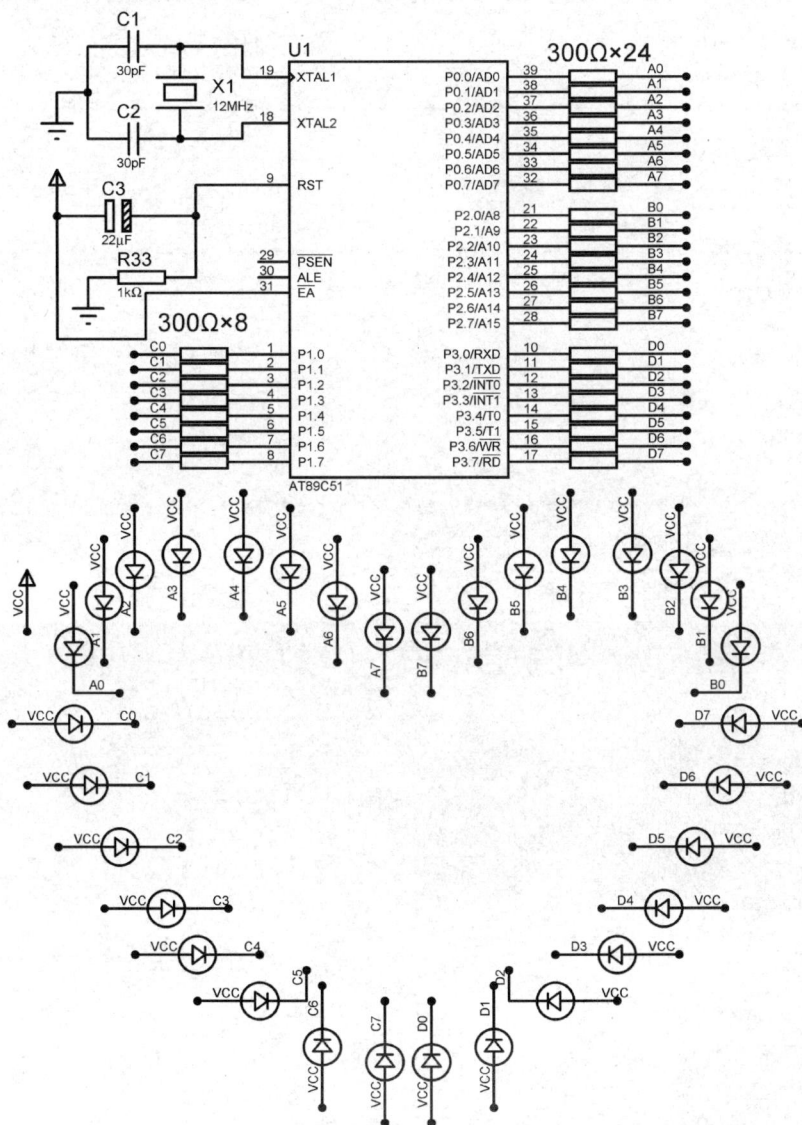

图 2-10 心形流水灯仿真图

3. 编写程序

采用移位的方式可以实现心形流水灯的顺时针点亮效果。

在 Keil μVision5 中编写程序，实现心形流水灯的效果，参考程序如下：

心形流水灯参考程序

```
/***********************************************************************
程序名称：program2-4.c
程序功能：心形流水灯程序
***********************************************************************/
#include<reg52.h>                        //加载头文件
#include<intrins.h>                      //加载头文件
/***********************************************************************
数据类型定义
***********************************************************************/
#define uchar unsigned char              //定义无符号字符型
#define uint unsigned int                //定义无符号整型
/***********************************************************************
函数名称：延时函数
功能描述：延时 t ms（晶振频率为 12MHz）
入口参数：t
***********************************************************************/
void delayms(uint t)
{
    uchar i;
    while(t--)
        for(i=96;i>0;i--);               //实现 1ms 延时
}
/***********************************************************************
函数名称：花式 1 显示函数
功能描述：实现信号灯的显示
入口参数：无
***********************************************************************/
void huashi_1()                          //花式 1，顺时针点亮
{
    uchar i=0xfe, j=0x7f, k;
    P0=0xff;                             //P0 口的 LED 全灭
    P1=0xff;                             //P1 口的 LED 全灭
    P3=0xff;                             //P3 口的 LED 全灭
    for(k=8;k>0;k--)
    {
        P2=j;                            //点亮 P2 口的 LED
        delayms(150);                    //延时 150ms
        j=_cror_(j,1);                   //P2 口的 LED 右移
    }
    P0=0xff;                             //P0 口的 LED 全灭
    P1=0xff;                             //P1 口的 LED 全灭
```

```
    P2=0xff;                              //P2 口的 LED 全灭
    for(k=8;k>0;k--)
    {
        P3=j;                             //点亮 P3 口的 LED
        delayms(150);                     //延时 150ms
        j=_cror_(j,1);                    //P2 口的 LED 右移
    }
    P0=0xff;                              //P0 口的 LED 全灭
    P3=0xff;                              //P3 口的 LED 全灭
    P2=0xff;                              //P2 口的 LED 全灭
    for(k=8;k>0;k--)
    {
        P1=j;                             //点亮 P1 口的 LED
        delayms(150);                     //延时 150ms
        j=_cror_(j,1);                    //P2 口的 LED 右移
    }
    P1=0xff;                              //P1 口的 LED 全灭
    P3=0xff;                              //P3 口的 LED 全灭
    P2=0xff;                              //P2 口的 LED 全灭
    for(k=8;k>0;k--)
    {
        P0=i;                             //点亮 P0 口的 LED
        delayms(150);                     //延时 150ms
        i=_crol_(I,1);                    //P0 口的 LED 左移
    }
}
/***************************************************************************
主程序
***************************************************************************/
void main()
{
    while(1)
    {
        huashi_1();                       //调用花式 1 显示函数
    }
}
```

4. 系统仿真

当 Keil C51 编译成功后，会自动产生 HEX 文件，接着打开之前绘制的
Proteus 仿真图，双击 AT89C51，弹出"Edit Component"对话框，单击"Program
File"中的文件夹按钮，在弹出的"Select File Name"对话框中，选择之前编
译生成的 HEX 文件，单击"打开"按钮，返回"Edit Component"对话框，单
击"OK"按钮，即可装入 HEX 文件。

心形流水灯仿
真效果视频

接着单击 Proteus ISIS 编辑界面左下角的运行按钮 ▶，即可观察是否能够实现心形流

水灯的顺时针点亮效果，如图 2-11 所示。

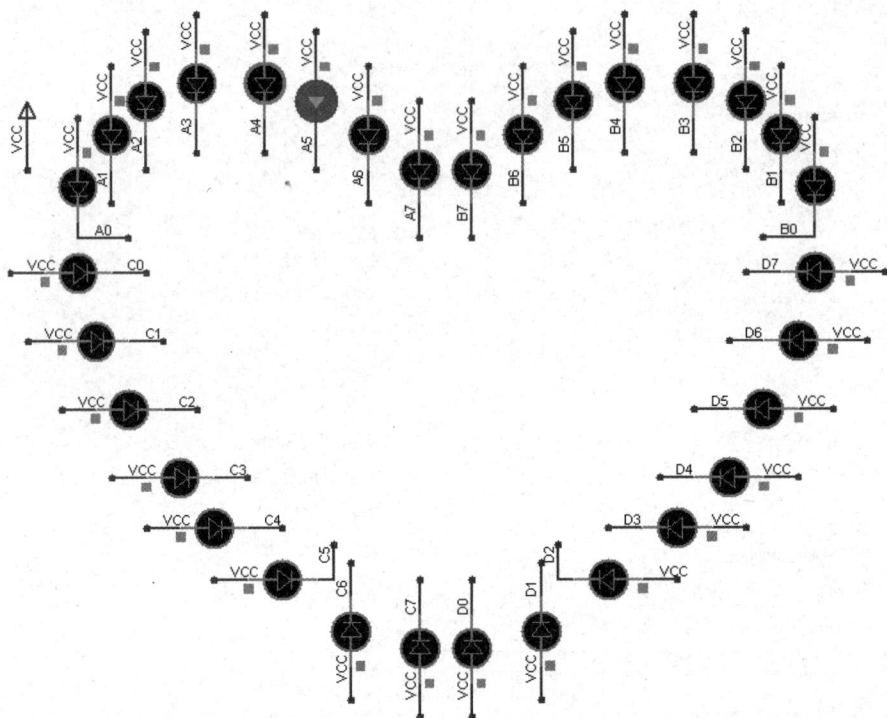

图 2-11　心形流水灯仿真效果图

素养小课堂

单片机的应用领域

单片机作为一种微型计算机，具有体积小、功耗低、功能强大等特点，被广泛应用于多个领域。以下是单片机的主要应用领域。

（1）智能化家用电器。

如今，几乎所有的现代家用电器都采用单片机进行智能化控制，包括但不限于洗衣机、空调、电视机、微波炉、电冰箱、电饭煲等。通过单片机，这些家用电器能够实现更加精准和智能的控制，提高用户的使用体验。

（2）办公自动化设备。

在现代办公室中，大量的通信和办公设备都嵌入了单片机。例如，打印机、复印机、传真机、绘图机、考勤机及通用计算机中的键盘译码、磁盘驱动等，都离不开单片机的控制。单片机使得这些设备能够高效地协同工作，提高办公效率。

（3）工业自动化控制。

工业自动化控制是单片机应用的一个重要领域。单片机可以构成各种工业控制系统和数据采集系统，实现生产过程的自动化和智能化。在数控机床、自动生产线控制、电机控制、温度控制等方面，单片机都发挥着核心作用。

（4）智能化仪器仪表。

单片机在仪器仪表领域也有广泛应用，如智能仪器、医疗器械、数字示波器等。单片机可以实现对各种物理量的测量和控制，提高仪器仪表的精度和稳定性。同时，单片机可以实现数据的处理和存储，为后续的数据分析提供有力支持。

（5）商用产品领域。

在商用产品领域，单片机也发挥着重要作用。例如，自动售货机、电子收款机、电子秤等商用设备都采用单片机进行控制。单片机使得这些设备能够实现自动化和智能化管理，提高商业运营效率和服务质量。

（6）汽车电子领域。

随着汽车电子化的不断发展，单片机在汽车电子领域的应用也越来越广泛。现代汽车的集中显示系统、动力监测控制系统、自动驾驶系统、通信系统和运行监视器等都离不开单片机的控制。单片机可以获取多种传感器采集的数据，并根据内部程序计算出数据的结果，从而控制汽车的行驶状态。

（7）航空航天和国防军事领域。

在航空航天和国防军事领域，单片机的应用同样重要。这些领域对设备的可靠性和精度要求极高，单片机凭借其强大的控制能力和稳定性，成为这些领域不可或缺的一部分。

综上所述，单片机在智能化家用电器、办公自动化设备、工业自动化控制、智能化仪器仪表、商用产品领域、汽车电子领域及航空航天和国防军事领域等多个方面都有广泛应用。随着技术的不断进步和需求的不断增加，单片机的应用领域还将不断扩展和深化。

课后任务

1. 修改任务 2.1 的设计要求，使得八路流水灯可以实现更多的显示效果（显示效果参见二维码）。

2. 修改任务 2.2 的设计要求，使得心形流水灯可以实现更多的显示效果（显示效果参见二维码）。

课后任务 1 仿真效果视频　　　　课后任务 2 仿真效果视频

知识拓展　单片机烧写器

1. 为什么称为"烧写"？

早期一般将调试好的单片机程序写入 ROM、EPROM，这种操作就像刻制光盘一样，是在高电压方式下写入的，PROM 是一次性写入的，存储器内部发生变化，有些线路或元器件会被烧断，不可恢复，所以称为烧写。而 EPROM 可以使用紫外线将原来写入的内容擦除，重新烧写。目前单片机大量采用的 EERPOM 或 FLASH ROM，是一种电可擦除的存储器。

2. 烧写的两种方式

（1）把单片机看作一个 ROM 芯片，早期的单片机都是如此。将单片机放在通用编程器上编程时，就像给 28C256 这样的 ROM 写入程序一样，只是不同的单片机使用的端口不同。图 2-12 所示为一种通用编程器 G540，这种通用编程器既可以烧写各种型号的单片机程序，又可以烧写 EEPROM。

（2）支持在线编程，如 STC 单片机、AVR 单片机、PIC 单片机等，可以使用专用的下载器，并将程序烧写到单片机中。图 2-13 所示为一款 PIC 单片机的在线编程器。

图 2-12　一种通用编程器 G540　　　　图 2-13　一款 PIC 单片机的在线编程器

3. STC 系列单片机烧录软件的使用

（1）STC 系列单片机烧录软件。

STC 系列单片机具有 ISP 技术，ISP 技术的优点：省去购买通用编程器，单片机在用户系统上即可下载/烧录用户程序，而不需要将单片机从已生产好的产品上拆下，再用通用编程器将程序烧录进单片机内部。有些程序尚未定型的产品可以一边生产，一边完善，加快产品进入市场的速度，减少新产品由于软件缺陷带来的风险。由于可以在用户的目标系统上将程序直接烧录进单片机看运行结果对错，故不需要仿真器。

STC 系列单片机的内部固化有 ISP 系统引导固件，配合 PC 端的控制程序即可将用户程序代码烧录进单片机内部，故不需要编程器（速度比通用编程器速度快，几秒一片）。

STC 提供的 ISP 下载工具可以从宏晶公司网站下载，支持 bin 和 HEX 文件，如果少数 HEX 文件不支持，则可转换成 bin 文件。

（2）STC 系列单片机的烧录步骤。

STC-ISP 烧录工具的完整界面如图 2-14 所示，烧录步骤如下。

步骤 1：选择单片机型号，如 STC89C52、STC12C2052AD 等。

步骤 2：选择串口，如串口 1—COM1、串口 2—COM2 等。

步骤 3：打开程序文件，要想烧录用户程序，必须调入用户程序代码。

步骤 4：单击"下载/编程"按钮，下载用户程序到单片机内部，可以重复执行步骤 4，也可以单击"重复编程"按钮。

下载时注意看提示，看是否要给单片机上电或复位，其下载速度比一般通用编程器速度快。

注意：一定要先单击"下载/编程"按钮，再给单片机上电复位（先彻底断电），而不要先上电，如果先上电，检测不到合法的下载命令流，单片机就直接运行用户程序了。

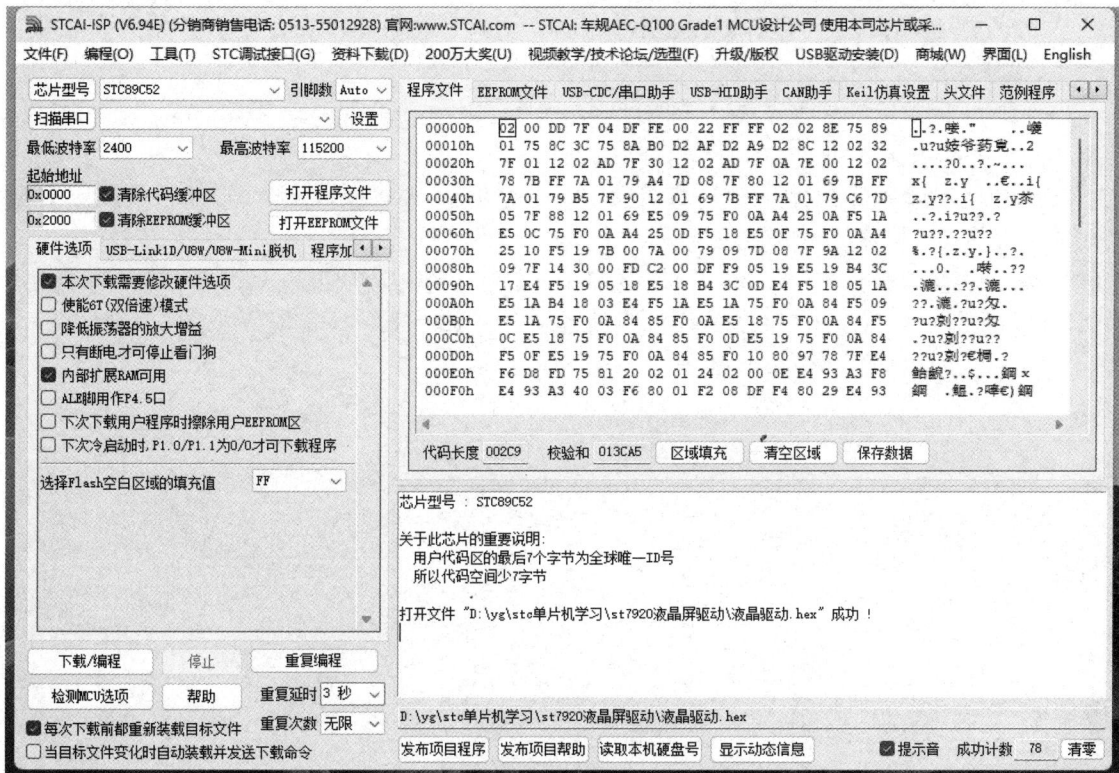

图 2-14　STC-ISP 烧录工具的完整界面

习题

一、单选题

1．Intel 8051 单片机内部包含（　　）个定时器/计数器。

A．1　　　　　　　　B．2　　　　　　　　C．3　　　　　　　　D．4

2．T0 可分为（　　）两部分。

A．TH0、TL0　　　C．TH1、TL1　　　B．TH0、TL1　　　D．TH1、TL0

3．数组若定义为"int a[5]={0,1,2,3};"，则下列说法错误的是（　　）。

A．a[0]=0　　　　　B．a[2]=2　　　　　C．a[5]=4　　　　　D．a[4]=0

4．（　　）是 51 单片机内部 T1 的启动控制位。

A．TR0　　　　　　B．TR1　　　　　　C．TF0　　　　　　D．TF1

5．当 M1 和 M0 为（　　）时，定时器/计数器被设置为方式 0。

A．01　　　　　　　B．10　　　　　　　C．00　　　　　　　D．11

6．设 8051 单片机的晶振频率为 12MHz，若用 T0 的方式 1 产生 1ms 定时，则 T0 计数初始值应为（　　）。

A．0xfc18　　　　　B．0xf830　　　　　C．0xf448　　　　　D．0xf060

7．8051 单片机的 T1 用作定时方式且工作在方式 1 时，工作方式控制字为（　　）。

A．TCON=0x01　　　B．TCON=0x02　　　C．TMOD=0x10　　　D．TMOD=0x50

二、编程题

1．编程实现单片机 P1.0 口输出周期为 20ms 的方波，要求采用定时器 T0 的方式 1 实现，已知单片机的晶振频率为 12MHz。

2．编程实现单片机 P1.0 口输出周期为 200μs 的方波，要求采用定时器 T1 的方式 2 实现，已知单片机的晶振频率为 12MHz。

3．编程实现单片机 P1.0 口输出周期为 10ms 的矩形波，矩形波占空比为 25%，即高电平为 2.5ms，低电平为 7.5ms，已知单片机的晶振频率为 12MHz。

项目三 时钟系统的仿真设计

任务 3.1 秒表系统的仿真设计

学习目标

【知识目标】

（1）了解并掌握数码管静态显示的使用方法。

（2）了解并掌握 C51 中 if 语句的使用方法。

秒表系统导学材料

【技能目标】

（1）了解并掌握单片机仿真软件 Proteus 的使用方法。

（2）了解并掌握单片机编译软件 Keil C51 的使用方法。

（3）了解并掌握单片机程序下载的方法。

（4）了解并掌握单片机最小系统的组成。

（5）通过秒表系统的仿真设计初步了解并掌握单片机项目的开发步骤。

【思政目标】

通过秒表系统的仿真设计，引导学生树立珍惜时间的观念，合理规划时间，高效利用每一分每一秒，将时间投入到有意义的学习和实践中去。

数码管静态显示
的实现微课视频

3.1.1 数码管静态显示的实现

1. LED 数码管显示器的结构

LED 数码管显示器是由发光二极管显示字段组成的显示器件，简称"数码管"。其外形结构如图 3-1（a）所示，由图 3-1 可知它由 8 个发光二极管（以下简称"字段"）构成，可通过不同的组合显示 0～9、A～F 及小数点等字符。图中的 dp 表示小数点，com 表示公共端。

数码管通常有共阴极和共阳极两种型号，如图 3-1（b）和图 3-1（c）所示。共阴极数码管内部所有发光二极管的阴极必须接低电平（一般为地），当某一发光二极管的阳极接高电平时，此二极管点亮；共阳极数码管内部所有发光二极管的阳极必须接高电平（一般为+5V），当某一发光二极管的阴极接低电平时，此二极管点亮。显然，要显示某字符就应使此字符的相应字段点亮，实际上就是送一个用不同电平组合代表的数据到数码管。这种装入数码管中显示字符的数据称为"字符编码"。

2. 数码管的字符编码

要使数码管显示出相应的字符，必须使数据口输出相应的字符编码。字符编码与硬件电路的连接形式有关。一般情况下将数码管数据口的低位和 a 相连，高位和 dp 相连，则数码管的字符编码的各位定义如表 3-1 所示。

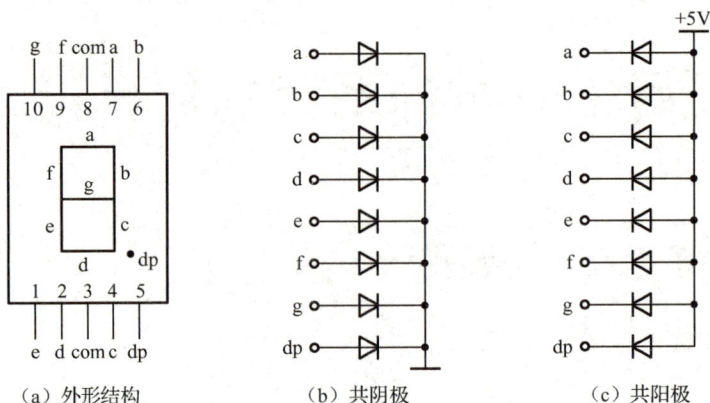

（a）外形结构　　　（b）共阴极　　　（c）共阳极

图 3-1　数码管结构图

表 3-1　数码管的字符编码的各位定义

位序	D7	D6	D5	D4	D3	D2	D1	D0
显示字段	dp	g	f	e	d	c	b	a

图 3-2　数字"0"的显示图

下面以数码管显示数字"0"为例，说明字符编码的方法。数字"0"的显示图如图 3-2 所示，其中 a、b、c、d、e、f 字段点亮，而 g 和 dp 字段不亮。

（1）共阴编码。

若采用共阴极数码管，则要点亮的字段为高电平，不亮的字段为低电平，数码管各字段的状态如表 3-2 所示，因此共阴极数码管显示数字"0"的十六进制编码为 3FH。

（2）共阳编码。

若采用共阳极数码管，则要点亮的字段为低电平，不亮的字段为高电平，数码管各字段的状态如表 3-3 所示，因此共阳极数码管显示数字"0"的十六进制编码为 C0H。

表 3-2　共阴极数码管各字段的状态

显示字段	dp	g	f	e	d	c	b	a
各字段状态	0	0	1	1	1	1	1	1

表 3-3　共阳极数码管各字段的状态

显示字段	dp	g	f	e	d	c	b	a
各字段状态	1	1	0	0	0	0	0	0

以此类推，可得到数码管的常用字符编码，如表 3-4 所示。

表 3-4　数码管的常用字符编码

字符	共阴极数码管编码									共阳极数码管编码								
	dp	g	f	e	d	c	b	a	编码	dp	g	f	e	d	c	b	a	编码
0	0	0	1	1	1	1	1	1	3FH	1	1	0	0	0	0	0	0	C0H
1	0	0	0	0	0	1	1	0	06H	1	1	1	1	1	0	0	1	F9H
2	0	1	0	1	1	0	1	1	5BH	1	0	1	0	0	1	0	0	A4H
3	0	1	0	0	1	1	1	1	4FH	1	0	1	1	0	0	0	0	B0H
4	0	1	1	0	0	1	1	0	66H	1	0	0	1	1	0	0	1	99H
5	0	1	1	0	1	1	0	1	6DH	1	0	0	1	0	0	1	0	92H
6	0	1	1	1	1	1	0	1	7DH	1	0	0	0	0	0	1	0	82H
7	0	0	0	0	0	1	1	1	07H	1	1	1	1	1	0	0	0	F8H
8	0	1	1	1	1	1	1	1	7FH	1	0	0	0	0	0	0	0	80H
9	0	1	1	0	1	1	1	1	6FH	1	0	0	1	0	0	0	0	90H
A	0	1	1	1	0	1	1	1	77H	1	0	0	0	1	0	0	0	88H
B	0	1	1	1	1	1	0	0	7CH	1	0	0	0	0	0	1	1	83H
C	0	0	1	1	1	0	0	1	39H	1	1	0	0	0	1	1	0	C6H
D	0	1	0	1	1	1	1	0	5EH	1	0	1	0	0	0	0	1	A1H
E	0	1	1	1	1	0	0	1	79H	1	0	0	0	0	1	1	0	86H
F	0	1	1	1	0	0	0	1	71H	1	0	0	0	1	1	1	0	8EH
R	0	1	1	1	0	0	0	0	31H	1	0	0	0	1	1	1	1	CEH
U	0	0	1	1	1	1	1	0	3EH	1	1	0	0	0	0	0	1	C1H
Y	0	1	1	0	1	1	1	0	6EH	1	0	0	1	0	0	0	1	91H
H	0	1	1	1	0	1	1	0	76H	1	0	0	0	1	0	0	1	89H
L	0	0	1	1	1	0	0	0	38H	1	1	0	0	0	1	1	1	C7H
-	0	1	0	0	0	0	0	0	40H	1	0	1	1	1	1	1	1	BFH
.	1	0	0	0	0	0	0	0	80H	0	1	1	1	1	1	1	1	7FH
全亮	1	1	1	1	1	1	1	1	FFH	0	0	0	0	0	0	0	0	00H
全灭	0	0	0	0	0	0	0	0	00H	1	1	1	1	1	1	1	1	FFH

3. 数码管的静态显示方式

静态显示是指当数码管显示某个字符时，相应的字段一直导通或截止，直到变换为其他字符。当数码管工作在静态显示方式下时，其公共端接地或高电平，每位的段选线与一个 8 位并行接口相连。只要在该位的段选线上保持段选码电平，该位就能保持相应的显示字符。这里的 8 位并行接口可以采用并行 I/O 接口（单片机的 I/O 接口或并行 I/O 接口芯片，如 8155、8255 芯片等），也可以采用串行输入/并行输出的移位寄存器。

静态显示的优点是稳定，在发光二极管的导通电流一定的情况下，显示器的亮度大，系统在运行时，仅仅在需要更新显示内容时 CPU 才执行一次显示子程序，这样大大节省了 CPU 时间，提高了 CPU 的工作效率；缺点是位数较多时显示口随之增加。

【课堂实训】采用单片机的 P2 口驱动一位共阳极数码管，要求数码管依次循环显示 0～9 十个数字，每个数字的显示时间为 1s。

按照题目要求，数码管与单片机的连接方式如图 3-3 所示。

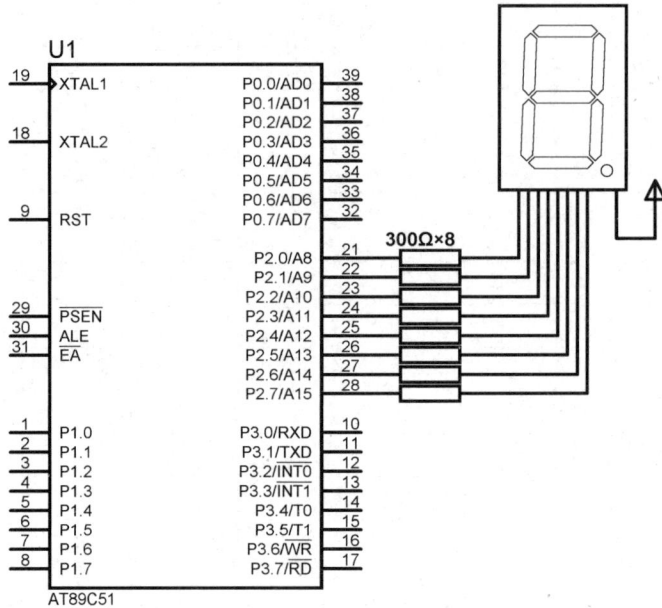

图 3-3　数码管与单片机的连接方式

参考程序如下：

```
/************************************************************
程序名称：program3-1.c
程序功能：P2 口驱动一位共阳极数码管
************************************************************/
#include<reg52.h>                        //加载头文件
/************************************************************
数据类型定义
************************************************************/
#define uchar unsigned char              //定义无符号字符型
#define uint unsigned int                //定义无符号整型
/************************************************************
数组定义：共阳极数码管 0～9 字符编码
************************************************************/
uchar code LED7SEG_A[]=
{    //0    1    2    3    4    5    6    7    8    9
     0xc0, 0xf9, 0xa4, 0xb0, 0x99, 0x92, 0x82, 0xf8, 0x80, 0x90
};
/************************************************************
函数名称：延时函数
功能描述：延时 t ms 函数（晶振频率为 12MHz）
```

```
入口参数：t
*******************************************************************/
void delayms(uint t)
{
    uchar i;
    while(t--)
        for(i=96;i>0;i--);
} /*****************************************************************
主函数
*******************************************************************/
void main()
{
    uchar i;
    while(1)
    {
        for(i=0;i<10;i++)
        {
            P2=LED7SEG_A[i];            //通过 P2 口送显示
            delayms(1000);             //延时 1s
        }
    }
}
```

3.1.2 C51 的 if 语句的使用

选择语句又称分支程序，判定给定的条件是否满足，根据判定结果（真或假）决定执行给出的操作程序，也被称为选择结构程序。在 C51 中，一般采用 if 语句或 switch 语句来实现。

1. 基本 if 语句

基本 if 语句的一般形式如下：

```
if(表达式)
{
    语句组;
}
```

C51 if 语句的使用微课视频

基本 if 语句的功能：如果表达式的结果为真，则执行花括号内的语句组，否则，跳过语句组执行下面的语句。

基本 if 语句的执行流程图如图 3-4 所示。

基本 if 语句中的表达式可以是任何形式的表达式，常用的是逻辑表达式或关系表达式。只要表达式的结果为真（非0），即可执行花括号内的语句组。

例 1：使用 if 语句编程实现，如果变量 a 大于 0，则变量 b 加 1。

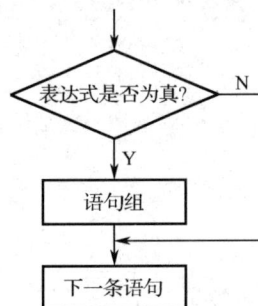

图 3-4 基本 if 语句的执行流程图

```
if(a>0)
{
    b++;
}
```

2. if-else 语句

基本 if 语句只在条件为真时执行指定的操作。而 if-else 语句为双分支 if 语句，在条件为真或为假时都有要执行的操作。

if-else 语句的一般形式如下：

```
if(表达式)
{
    语句组 1;
}
else
{
    语句组 2;
}
```

if-else 语句的功能：如果表达式的结果为真，则执行语句组 1；如果表达式的结果为假，则执行语句组 2。if-else 语句的执行流程图如图 3-5 所示。

例 2：使用 if-else 语句编程实现，如果单片机 P1.0 引脚为低电平，则 P2 口的值为 0x55，否则 P2 口的值为 0xaa。

```
sbit IN=P1.0;        //定义 P1.0 引脚
if(IN==0)            //判断 P1.0 引脚是否为低电平
    P2=0x55;         //若是，则 P2 口的值为 0x55
else
    P2=0xaa;         //若否，则 P2 口的值为 0xaa
```

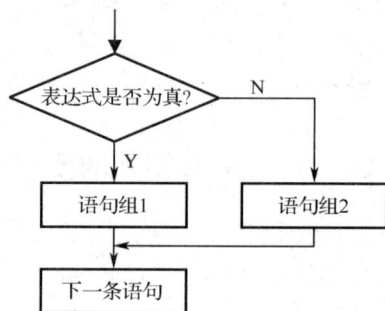

图 3-5　if-else 语句的执行流程图

3. if-else-if 语句

if-else-if 语句是一种多分支 if 语句，其一般形式如下：

```
if(表达式 1)
{
    语句组 1;
}
else if(表达式 2)
{
    语句组 2;
}
…
else if(表达式 n)
{
    语句组 n;
```

```
}
else
{
    语句组 n+1;
}
```

从 if-else-if 语句的形式中可以看出，它是 if-else 语句的嵌套形式，构成多分支的选择结构，其执行流程图如图 3-6 所示。

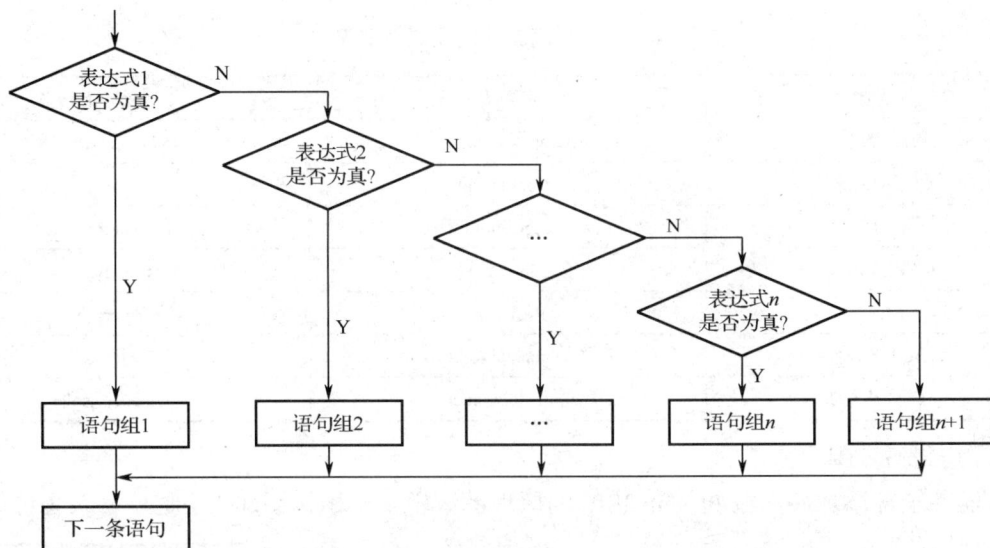

图 3-6　if-else-if 语句的执行流程图

例 3：使用 if-else-if 语句编程实现，若单片机 P1 口的值小于 0x30，则 P3 口的值为 0x55；若 P1 口的值大于或等于 0x30，且小于 0xa0，则 P3 口的值为 0xaa；若 P1 口的值大于或等于 0xa0，且小于 0xc0，则 P3 口的值为 0x5a；若 P1 口的值大于或等于 0xc0，则 P3 口的值为 0xa5。

```
if(P1<0x30)              //判断 P1 口的值是否小于 0x30
    P3=0x55;             //若是，则 P3 口的值为 0x55
else if(P1<0xa0)         //判断 P1 口的值是否大于或等于 0x30，且小于 0xa0
    P3=0xaa;             //若是，则 P3 口的值为 0xaa
else if(P1<0xc0)         //判断 P1 口的值是否大于或等于 0xa0，且小于 0xc0
    P3=0x5a;             //若是，则 P3 口的值为 0x5a
else                     //判断 P1 口的值是否大于或等于 0xc0
    P3=0xa5;             //若是，则 P3 口的值为 0xa5
```

3.1.3　秒表系统的任务实施

【设计要求】

通过单片机设计一个秒表，采用 2 个共阳极数码管显示秒表时间，显示范围为 0～59，每秒显示加 1，循环显示。

【任务分析】

根据设计要求，单片机需要在 P2 口输出秒表的十位数，在 P3 口输出秒表的个位数。因此，将秒表的值除以 10 可以得到秒表的十位数，将秒表的值对 10 取余，可以得到秒表的个位数，然后查共阳极数码管编码表输出相应的编码给 LED 数码管即可实现显示。

【实施步骤】

1. 添加元器件

打开 Proteus 仿真软件，按照表 3-5 添加元器件。

表 3-5　秒表系统的元器件清单

元器件名称	所属类	所属子类
AT89C51	Microprocessor ICs	8051 Family
CRYSTAL	Miscellaneous	—
CAP	Capacitors	Generic
CAP-ELEC	Capacitors	Generic
RES	Resistors	Generic
BUTTON	Switches & Relays	Switches
7SEG-MPX1-CA	Optoelectronics	7-Segment Displays

2. 绘制仿真图

元器件全部添加后，在 Proteus ISIS 的原理图编辑窗口中按图 3-7 绘制秒表系统仿真图。

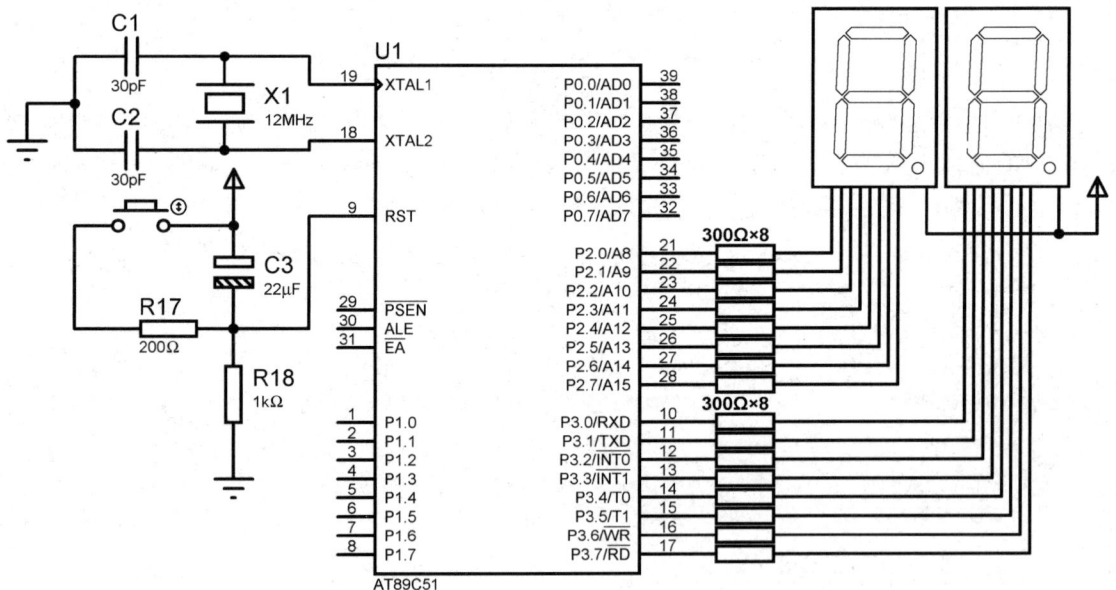

图 3-7　秒表系统仿真图

3. 编写程序

首先设计一个 60 次的 for 循环，循环主体为秒表数据显示和 1s 延时函数。采用一维数组的方式定义共阳极数码管 0～9 十个数字的编码表，然后通过查数组的方式实现数码管的显示。程序包含主程序、延时函数两部分。其中，主程序的流程图如图 3-8 所示。

图 3-8　主程序的流程图

秒表系统参考程序

在 Keil μVision5 中编写程序，实现秒表系统的效果，参考程序如下：

```
/*****************************************************
程序名称：program3-2.c
程序功能：秒表系统程序
*****************************************************/
#include<reg52.h>                        //加载头文件
/*****************************************************
数据类型定义
*****************************************************/
#define uchar unsigned char              //定义无符号字符型
#define uint unsigned int                //定义无符号整型
/*****************************************************
数组定义：共阳极数码管 0~9 十个数字的编码表
*****************************************************/
uchar code LED7SEG_A[]=
{     //0    1    2    3    4    5    6    7    8    9
    0xc0, 0xf9, 0xa4, 0xb0, 0x99, 0x92, 0x82, 0xf8, 0x80, 0x90
};
/*****************************************************
函数名称：延时函数
功能描述：延时 t ms（晶振频率为 12MHz）
入口参数：t
*****************************************************/
void delayms(uint t)
{
    uchar i;
    while(t--)
        for(i=96;i>0;i--);               //实现 1ms 延时
}
```

```
/*********************************************************************
主程序
*********************************************************************/
void main()                              //主函数
{
    uchar i;
    while(1)
    {
        for(i=0;i<60;i++)                //实现 60 次循环
        {
            P2=LED7SEG_A[i/10];          //输出秒表的十位数
            P3=LED7SEG_A[i%10];          //输出秒表的个位数
            delayms(1000);               //延时 1s
        }
    }
}
```

4. 系统仿真

当 Keil C51 编译成功后，会自动产生 HEX 文件，接着打开之前绘制的 Proteus 仿真图，双击 AT89C51，弹出"Edit Component"对话框，单击"Program File"中的文件夹按钮，在弹出的"Select File Name"对话框中，选择之前编译生成的 HEX 文件，单击"打开"按钮，返回"Edit Component"对话框，单击"OK"按钮，即可装入 HEX 文件。

秒表系统仿真效果视频

接着单击 Proteus ISIS 编辑界面左下角的运行按钮 ▶ ，即可观察是否能够实现秒表系统的显示效果，如图 3-9 所示。

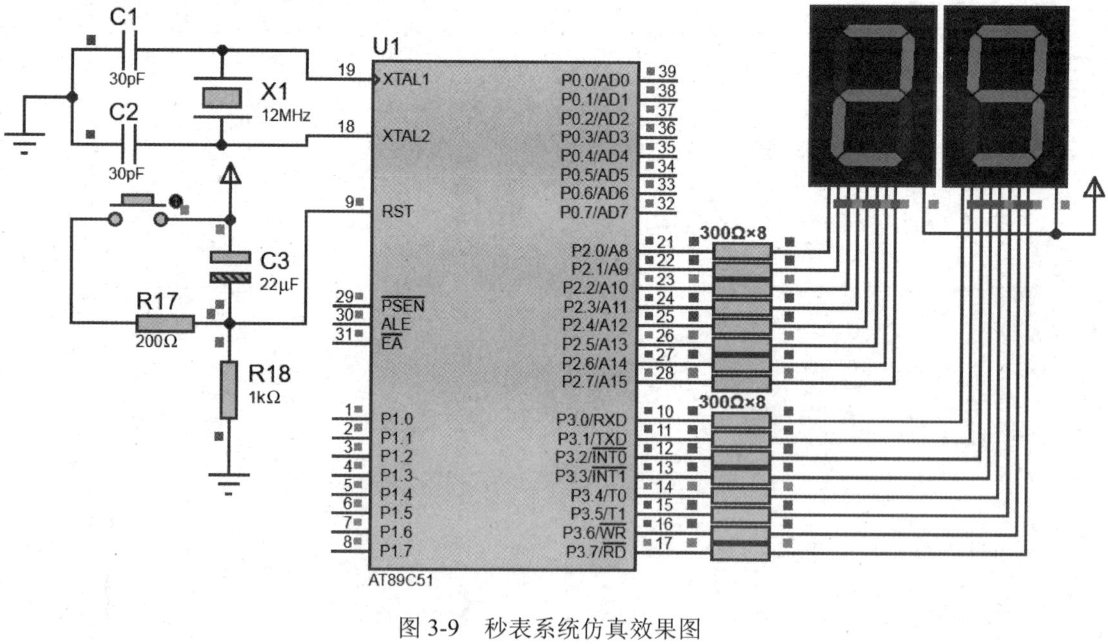

图 3-9　秒表系统仿真效果图

任务 3.2 倒计时系统的仿真设计

学习目标

【知识目标】

（1）了解并掌握数码管动态显示的原理。

（2）了解并掌握数码管动态显示的使用方法。

倒计时系统导学材料

【技能目标】

（1）了解并掌握单片机仿真软件 Proteus 的使用方法。

（2）了解并掌握单片机编译软件 Keil C51 的使用方法。

（3）通过倒计时系统的仿真设计进一步掌握单片机项目的开发步骤。

【思政目标】

通过讲解数码管动态扫描原理，让学生了解"眼见不一定为实"的哲学道理，学会辩证地分析问题，全面地分析和思考问题，从而提高辨识能力和社会责任意识。

3.2.1 数码管动态显示的实现

数码管动态显示方式是把各显示器的相同段选线并联在一起，并由一个 8 位 I/O 接口控制，而其公共端由其他相应的 I/O 接口控制，然后采用扫描方法轮流点亮各位 LED，使每位都分时显示该位应该显示的字符，这是最常用的显示方式之一。

数码管动态显示微课视频

在轮流点亮扫描的过程中，每位显示器的点亮时间是极为短暂的（约为 1ms），但由于人眼的视觉暂留现象及发光二极管的余辉效应，尽管实际上各位显示器并非同时显示，但只要扫描的速度足够快，给人的印象就是一组稳定的显示数据，不会有闪烁感。为了保证足够的亮度，通过 LED 的脉冲电流应是其额定电流的数倍。动态显示驱动电路是单片机应用系统中最常用的显示方式。图 3-10 所示为一种动态显示方式的示意图，其中数码管为共阳极数码管。

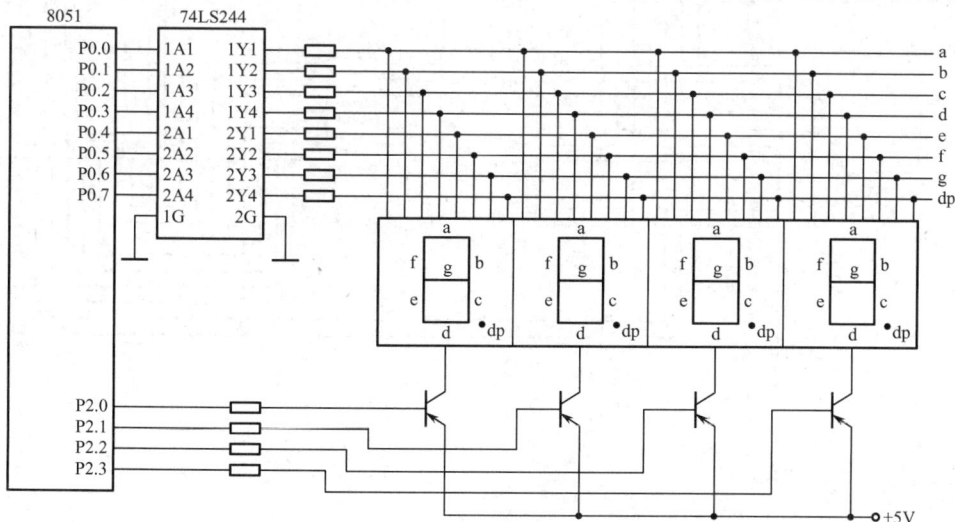

图 3-10 一种动态显示方式的示意图

3.2.2 倒计时系统的任务实施

【设计要求】

使用单片机设计一个倒计时系统，采用二位一体共阴极数码管显示倒计时时间，初始值为 20，每隔 1s 减 1，减为 0 后重新显示 20。

【任务分析】

在本任务中，单片机 P0 口输出二位一体共阴极数码管的段码，P2 口驱动数码管的位码端。

【实施步骤】

1. 添加元器件

打开 Proteus 仿真软件，按照表 3-6 添加元器件。注意：用 Proteus 仿真软件绘制单片机仿真图时，可以省略振荡电路和复位电路。

表 3-6　倒计时系统的元器件清单

元器件名称	所属类	所属子类
AT89C51	Microprocessor ICs	8051 Family
RES	Resistors	Generic
74LS245	TTL 74LS series	Transceivers
RESPACK-8	Resistors	Resistor Packs
7SEG-MPX2-CC	Optoelectronics	7-Segment Displays

2. 绘制仿真图

元器件全部添加后，在 Proteus ISIS 的原理图编辑窗口中按图 3-11 绘制倒计时系统仿真图。

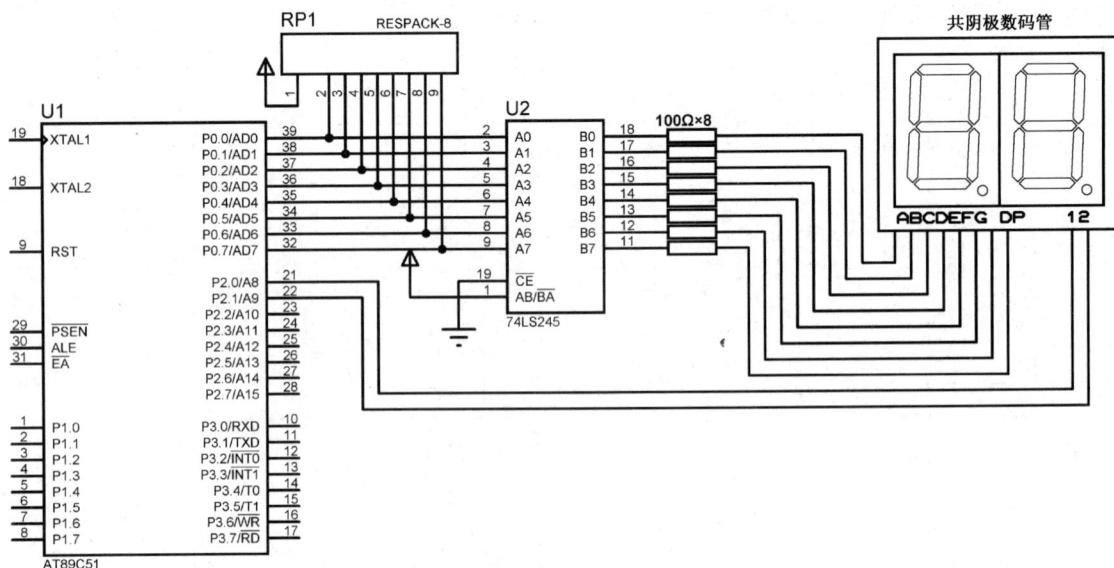

图 3-11　倒计时系统仿真图

3. 编写程序

在 Keil μVision5 中编写程序，实现倒计时系统的效果，参考程序如下：

```
/*************************************************
程序名称：program3-3.c
程序功能：倒计时系统程序
*************************************************/
#include<reg52.h>                      //加载头文件
#include<intrins.h>                    //加载头文件
/*************************************************
数据类型定义
*************************************************/
#define uchar unsigned char            //定义无符号字符型
#define uint unsigned int              //定义无符号整型
/*************************************************
全局变量定义
*************************************************/
uchar TIME=20;                         //倒计时初始值存储单元
uchar DISPLAY_DATA[2]={0};             //显示单元定义
/*************************************************
数组定义：共阴极数码管 0～9 十个数字的编码表
*************************************************/
uchar code LED_7SEG_CC[]=
{   //0     1     2     3     4     5     6     7     8     9
    0x3f, 0x06, 0x5b, 0x4f, 0x66, 0x6d, 0x7d, 0x07, 0x7f, 0x6f
};
/*************************************************
函数名称：延时函数
功能描述：延时 t ms（晶振频率为 12MHz）
入口参数：t
*************************************************/
void delayms(uint t)
{
    uchar i;
    while(t--)
        for(i=96;i>0;i--);             //实现 1ms 延时
}
/*************************************************
函数名称：显示函数
功能描述：共阴极数码管动态显示
```

```
入口参数：无
*******************************************************************/
void display()
{
    uchar i,j=0xfe;                          //共阴驱动
    for(i=0;i<2;i++)                         //共 2 个数码管
    {
        P2=j;                                //点亮一个数码管
        P0=LED_7SEG_CC[DISPLAY_DATA[i]];     //输出数据
        j=_crol_(j,1);                       //准备点亮下一个数码管
        delayms(1);                          //延时 1ms
    }
}
/*******************************************************************
主程序
*******************************************************************/
void main()
{
    uint i;
    while(1)
    {
        DISPLAY_DATA[0]=TIME/10;             //获得倒计时时间的十位数
        DISPLAY_DATA[1]=TIME%10;             //获得倒计时时间的个位数
        for(i=500;i>0;i--)                   //2ms×500=1s
            display();                       //调用显示函数，时间为 2ms
        TIME--;                              //倒计时时间减 1
        if(TIME==0xff)                       //倒计时时间减为 0 后，自动变为最大值
            TIME=20;
    }
}
```

4. 系统仿真

当 Keil C51 编译成功后，会自动产生 HEX 文件，接着打开之前绘制的 Proteus 仿真图，双击 AT89C51，弹出"Edit Component"对话框，单击"Program File"中的文件夹按钮，在弹出的"Select File Name"对话框中，选择之前编译生成的 HEX 文件，单击"打开"按钮，返回"Edit Component"对话框，单击"OK"按钮，即可装入 HEX 文件。

简易倒计时系统
仿真效果视频

接着单击 Proteus ISIS 编辑界面左下角的运行按钮 ▶，即可观察是否能够实现 20s 倒计时功能，如图 3-12 所示。

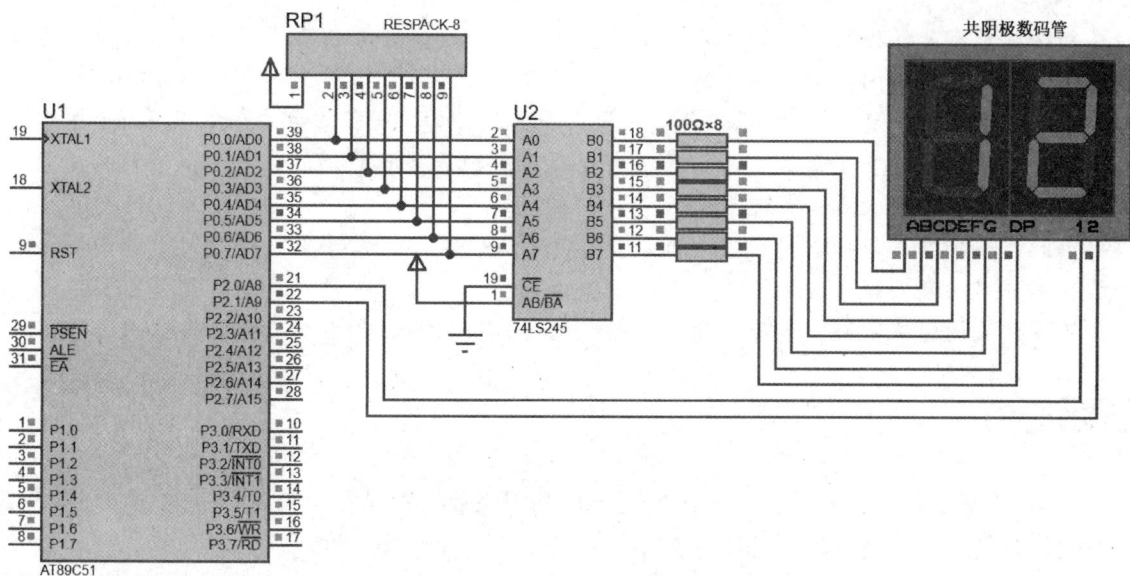

图 3-12　倒计时系统仿真效果图

素养小课堂

基于"眼见不一定为实"的数码管动态显示

在秒表设计任务中，时间的显示采用两位数码管实现，如显示"27"这两位数，是根据数码管动态显示的原理实现的。所谓动态显示，就是利用循环扫描的方式，分时轮流选通各数码管的公共端，使各数码管轮流导通工作。即"27"这两位数不是同时显示出来的，而是分时先显示"2"，再显示"7"。当计算机扫描速度足够快时，由于人眼有视觉暂留现象，就分辨不出来了，认为是各数码管同时显示，因此能够看见"27"，如图 3-13 所示。

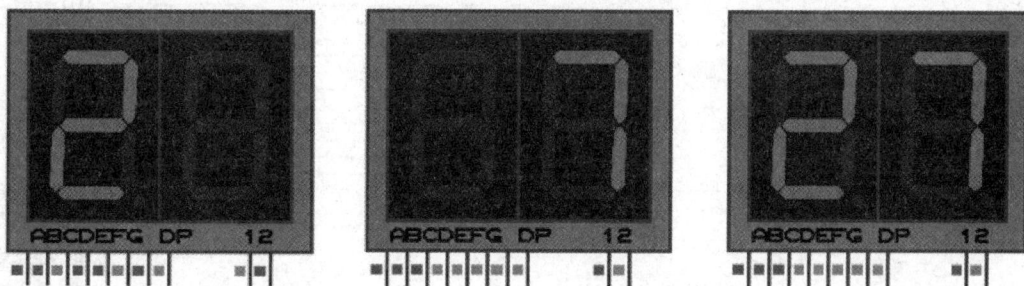

图 3-13　数码管动态显示的分解图

数码管动态扫描这一原理蕴含了"眼见不一定为实"的哲学道理，它挑战了我们对直接感知的信赖，明明是两位数码管分时工作，但我们的眼睛却看到两位数码管一起显示。这就说明了现象与本质的区别，我们的眼睛往往只能捕捉到现象，而无法直接洞察事物的本质。因此，"眼见"的往往只是现象，而非事物的本质。

人的感知能力是有限的，受到生理结构、心理状态、文化背景等多种因素的影响。我们的感知往往是有选择性的、有偏见的，甚至可能是错误的。因此，"眼见"的并不一定就是真实的，它可能只是我们主观感知的结果。

既然"眼见不一定为实"，那么我们需要采用更加科学、理性的方法来探求真理，这包括运用逻辑思维、科学实验、历史考证等多种手段来验证和修正我们的认知。只有这样，我们才能逐步接近事物的本质和客观真理。

因此，我们在思考知识与现实之间的复杂关系时，要多问、多看、多思考，辩证地分析问题，不要只看表面现象，人云亦云。特别是在对待网络上的一些新闻时，要多求证、多思考，不要被表面现象蒙蔽，学会全面地分析和思考，提高辨识能力和社会责任意识。

课后任务

1. 编程实现一个秒表系统，要求采用 3 位共阳极数码管显示，显示范围为 00.0~19.9s，并且重复循环显示，其电路图如图 3-14 所示，显示效果参见二维码。

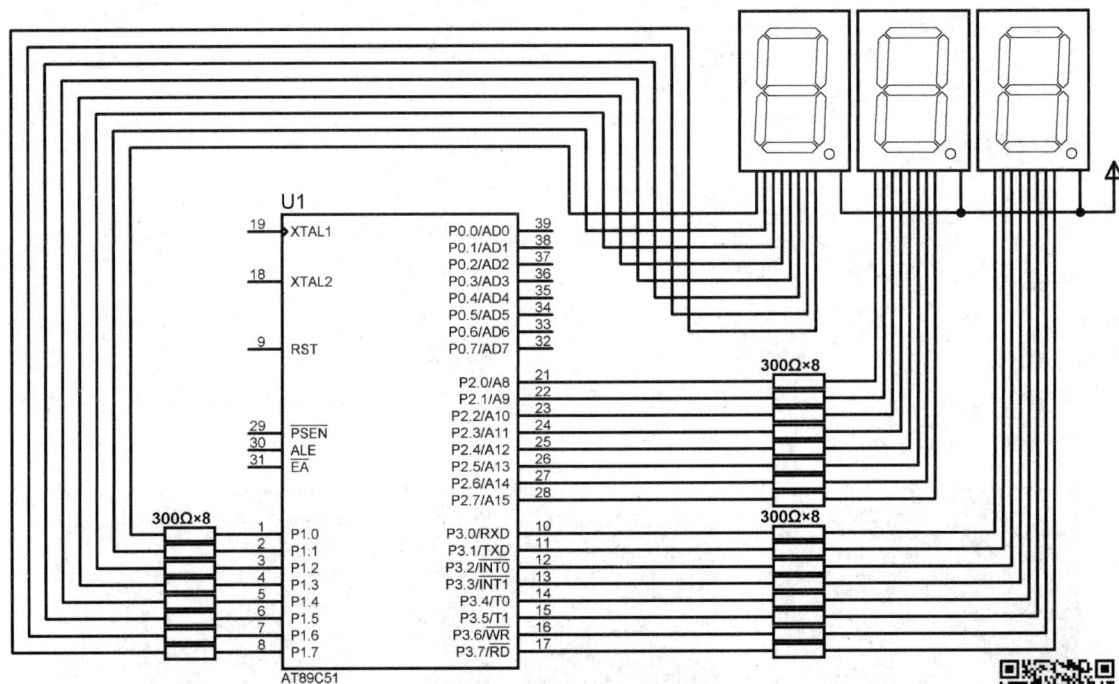

图 3-14　课后任务 1 电路图

课后任务 1
仿真效果视频

2. 编程实现一个计时系统，要求采用 4 位共阳极数码管显示，显示范围为 0 分 0 秒~59 分 59 秒，并且重复循环显示，其电路图如图 3-15 所示，显示效果参见二维码。

3. 编程实现一个 24 小时计时系统，要求采用八位一体共阳极数码管显示，显示范围为 0 时 0 分 0 秒~23 时 59 分 59 秒，并且重复循环显示。要求系统开机时，初始值为 23-59-40，其电路图如图 3-16 所示，显示效果参见二维码。

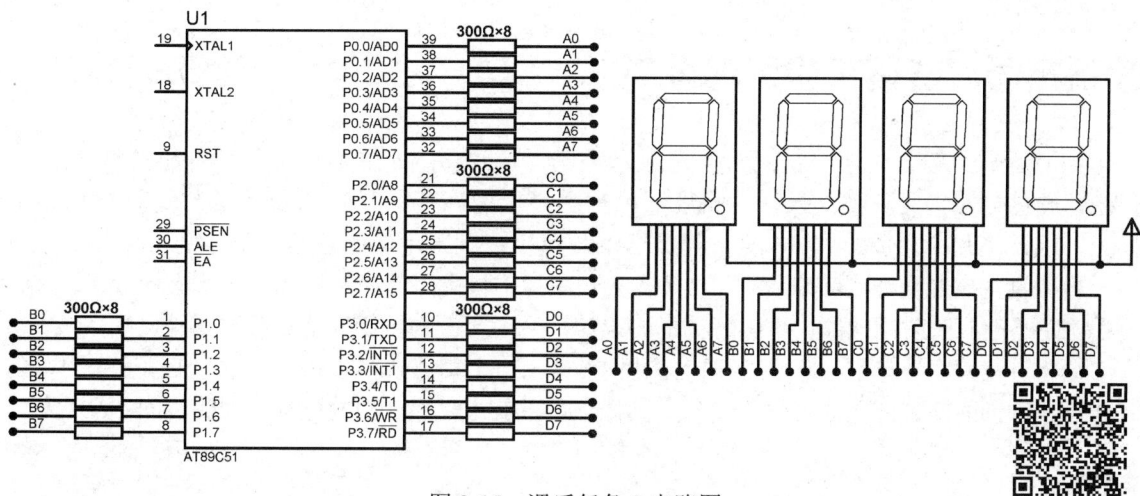

图 3-15　课后任务 2 电路图

课后任务 2
仿真效果视频

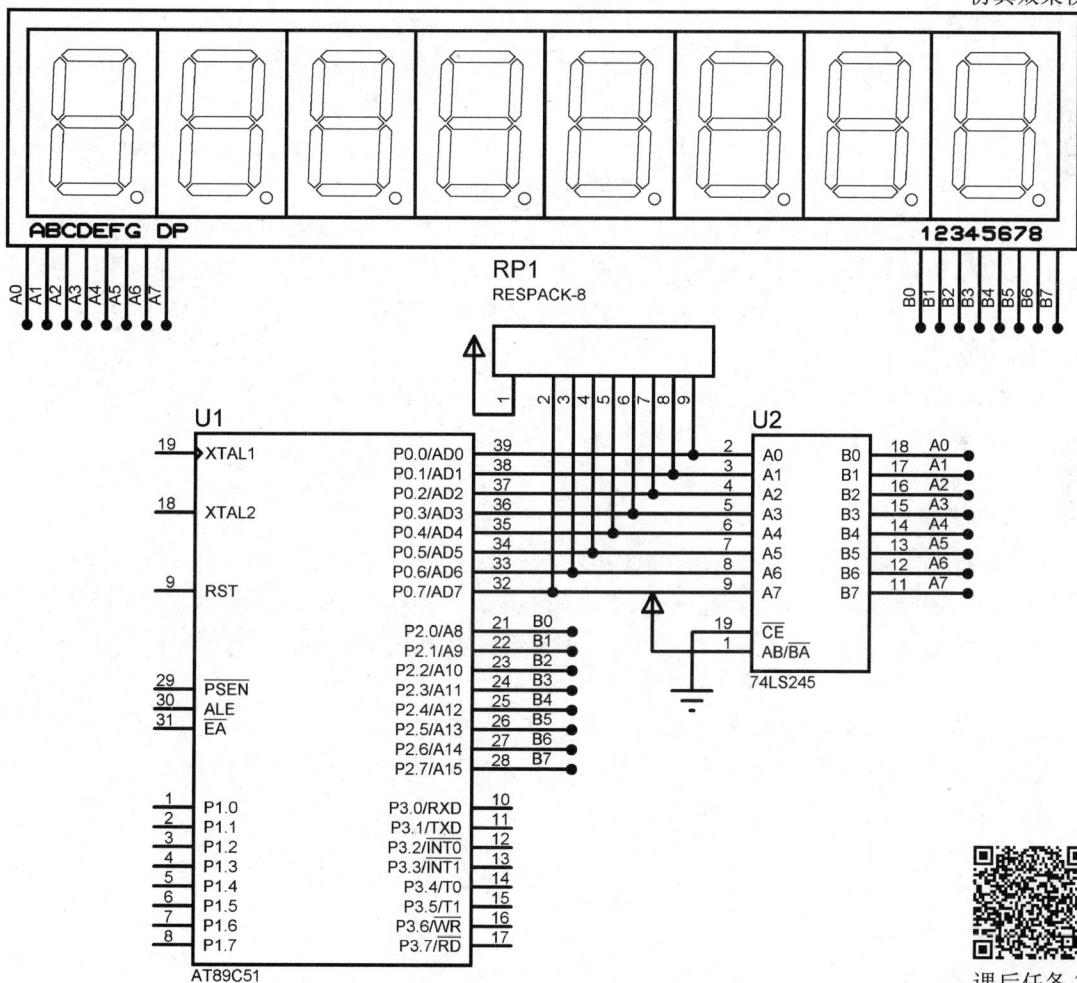

图 3-16　课后任务 3 电路图

课后任务 3
仿真效果视频

知识拓展　数码管显示的优点

当显示位数较多时，与数码管静态显示相比，数码管动态显示可节省单片机的 I/O 接口资源，硬件电路简单，但其显示的亮度低于静态显示的亮度。由于 CPU 不断地运行扫描显示程序，因此占用 CPU 更多的时间。若显示位数较少，则采用数码管静态显示更加简便。

数码管动态显示在实际应用中，由于需要不断地扫描数码管才能得到稳定的显示效果，因此在程序中不能有长时间地停止数码管扫描的语句，否则会影响显示效果，甚至无法显示。

通常，在程序设计中，把数码管扫描过程设计为一个相对独立的扫描函数，在程序中需要延时或等待查询的地方调用该函数，代替空操作延时，就可以保证扫描过程的间隔时间不会太长。

习题

一、单选题

1. 已知在共阴极数码管中，a 字段为字符编码的最低位。若需要显示字符 H，则它的字符编码应为（　　）。

A. 76H　　　　　　　B. 7FH　　　　　　　C. 80H　　　　　　　D. F6H

2. LED 数码管用作动态显示时，需要（　　）。

A. 将各位数码管的位码线并联起来　　　　B. 将各位数码管的位码线串联起来

C. 将各位数码管的相同段码线并联起来　　D. 将各位数码管的相同段码线串联起来

3. 已知 1 个共阴极数码管，其中 a 字段为字符编码的最低位，若需要显示数字 1，则它的字符编码应为（　　）。

A. 06H　　　　　　　B. F9H　　　　　　　C. 30H　　　　　　　D. CFH

4. 在 C 语言的 if 语句中，用作判断的表达式为（　　）。

A. 关系表达式　　　B. 逻辑表达式　　　C. 算术表达式　　　D. 任意表达式

二、多选题

1. 下列关于数码管动态显示的描述中，错误的是（　　）。

A. 一个并行接口只接一个数码管，显示数据送入并行接口后就不再需要 CPU 干预

B. 动态显示只能使用共阴极数码管，不能使用共阳极数码管

C. 数码管动态扫描时间一般设置为 1s

D. 动态显示具有占用 CPU 时间少、发光亮度稳定的特点

2. LED 数码管若采用动态显示方式，则下列说法中正确的是（　　）。

A. 将各位数码管的段选线并联

B. 将段选线用一个 8 位 I/O 接口控制

C. 将各位数码管的公共端直接接在+5V 或 GND 上

D. 将各位数码管的位选线用各自独立的 I/O 接口控制

三、判断题

1. 八段共阴极数码管显示数字 0 时的字段码是 0xc0。　　　　　　　　　　（　　）

2. 数码管动态扫描时间一般控制在 1μs。　　　　　　　　　　　　　　　　（　　）

项目四 抢答器系统的仿真设计

任务 4.1 八路抢答器的仿真设计

学习目标

【知识目标】

（1）了解并掌握独立按键的识别原理和使用方法。

（2）了解并掌握 C51 的 switch 语句的使用方法。

八路抢答器导学材料

【技能目标】

（1）了解并掌握单片机仿真软件 Proteus 的使用方法。

（2）了解并掌握单片机编译软件 Keil C51 的使用方法。

（3）了解并掌握单片机程序下载的方法。

（4）了解并掌握单片机最小系统的组成。

（5）通过八路抢答器的仿真设计初步了解并掌握单片机项目的开发步骤。

【思政目标】

通过八路抢答器的仿真设计，学生可以了解抢答过程中的公正裁决和透明操作，亲身体验到公平正义的价值，从而增强社会责任感和正义感。

4.1.1 独立按键的识别

独立按键的识别微课视频

1. 按键的结构和工作原理

在单片机应用系统中，除了复位键有专门的复位电路及专一的复位功能，其他按键均以开关状态来设置控制功能或输入数据，因此，这些按键只是简单的电平输入。按键信息输入是与软件功能密切相关的过程。对于某些应用系统，如智能仪表，按键输入程序是整个应用程序中的重要组成部分。

独立按键是指直接用 I/O 接口线构成的单个按键电路。每个独立按键都单独地占有一根 I/O 接口线，每根 I/O 接口线的工作状态都不会影响其他 I/O 接口线的工作状态，这是一种最简单、易懂的按键结构。

每个独立按键都是一个常开的开关电路，当所设置的功能键或数字键被按下时，处于闭合状态。图 4-1 所示为独立按键与单片机的连接电路，当 K 键未被按下时，由于上拉电阻 R1 的作用，单片机的 P3.7 口为高电平；当 K 键被按下时，P3.7 口为低电平。

2. 按键的抖动

按键大多为触点式按键，由于触点的弹性作用，当按键闭合时，不会马上稳定地接通；

当按键断开时，也不会立即断开。即在按键闭合、断开的瞬间，均伴随着一连串的抖动，抖动时间的长短由按键的机械特性决定，一般为 5~10ms。按键抖动示意图如图 4-2 所示。

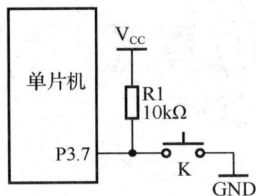

图 4-1　独立按键与单片机的连接电路

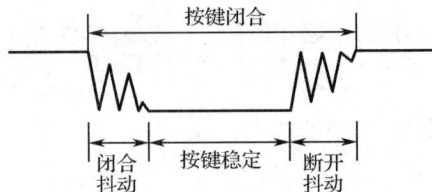

图 4-2　按键抖动示意图

图 4-3　按键去抖动的流程图

为了克服按键触点机械抖动所致的检测误判，确保一次按键动作只确认一次按键，必须采用去抖动措施。

去抖动措施主要有硬件去抖动和软件去抖动两种。在硬件上，采取在按键输出端加 RS 触发器或单稳态电路的方式构成去抖动电路。在软件上，检测到有按键被按下时，先执行一个 10ms 左右的延时程序，再判断该按键电平是否仍保持为闭合状态电平，若是，则该按键处于闭合状态，否则认为是干扰信号，从而去除了抖动影响。为简化电路，通常采用软件去抖动。按键去抖动的流程图如图 4-3 所示。

【课堂实训】编程实现按键控制数码管显示效果，单片机的 P1.0 口接一个独立按键，P2 口驱动共阳极数码管。要求系统复位时，数码管显示 0，每按下一次按键，数码管显示就加 1，当数码管显示 9 时，再按下一次按键，数码管重新显示 0。电路连接图如图 4-4 所示。

课堂实训
仿真效果视频

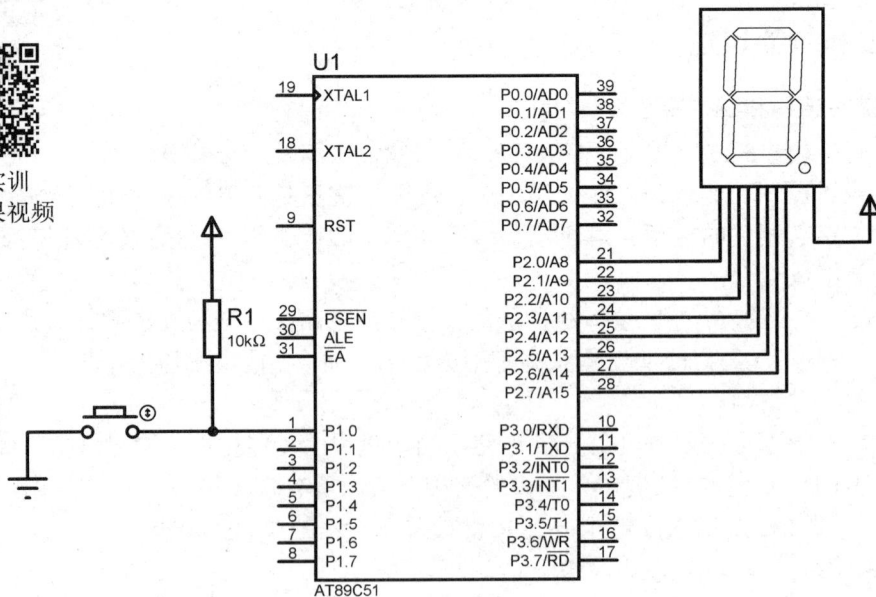

图 4-4　电路连接图

参考程序如下：

```
/*********************************************************************
程序名称：program4-1.c
程序功能：按键控制数码管显示效果
*********************************************************************/
#include<reg52.h>                          //加载头文件
/*********************************************************************
数据类型定义
*********************************************************************/
#define uchar unsigned char                //定义无符号字符型
/*********************************************************************
引脚定义
*********************************************************************/
sbit KEY=P1^0;                             //按键引脚定义
/*********************************************************************
全局变量定义
*********************************************************************/
uchar KEY_DATA=0;                          //按键值
/*********************************************************************
数组定义：共阳极数码管 0～9 十个数字的编码表
*********************************************************************/
uchar code LED7SEG_CA[]=
{    //0     1     2     3     4     5     6     7     8     9
    0xc0, 0xf9, 0xa4, 0xb0, 0x99, 0x92, 0x82, 0xf8, 0x80, 0x90
};
/*********************************************************************
函数名称：延时函数
功能描述：延时 t ms 函数（晶振频率为 12MHz）
入口参数：t
*********************************************************************/
void delayms(uint t)
{
    uchar i;
    while(t--)
        for(i=96;i>0;i--);
}
/*********************************************************************
函数名称：按键查询函数
*********************************************************************/
void key_scan()                            //按键查询函数
{
    if(KEY==0)                             //判断按键是否被按下
    {
        delayms(20);                       //延时 20ms 去抖动
```

```
            if(KEY==0)                          //再次判断按键是否被按下
            {
                while(KEY==0);                  //判断按键是否被释放
                KEY_DATA++;                     //若被释放，则按键值加 1
                if(KEY_DATA==10)                //判断按键值是否溢出
                    KEY_DATA=0;                 //若溢出，则清零
            }
        }
}
/****************************************************************
主函数
****************************************************************/
void main()
{
    while(1)
    {
        key_scan();                             //按键查询函数
        P2=LED7SEG_CA[KEY_DATA];                //显示
    }
}
```

4.1.2   C51 的 switch 语句的使用

C51 的 switch 语句的使用微课视频

if 语句一般用于单一条件或分支数目较少的场合，如果使用 if 语句编写具有三个以上分支的程序，就会降低程序的可读性。C 语言提供了一种用于多分支选择的 switch 语句，其一般形式如下：

```
switch(测试表达式)
{
    case   常量表达式 1: 语句 1;   break;
    case   常量表达式 2: 语句 2;   break;
    ...
    case   常量表达式 n: 语句 n;   break;
    default: 语句 n+1;
}
```

switch 语句的执行流程图如图 4-5 所示，首先计算测试表达式的值，并逐个与 case 语句后的常量表达式的值进行比较，当测试表达式的值与某个常量表达式的值相等时，执行该常量表达式后的语句，再执行 break 语句，跳出 switch 语句的执行，继续执行其他语句。如果测试表达式的值与所有 case 语句后的常量表达式的值均不相等，则执行 default 后的语句。

在使用 switch 语句时，需要注意以下情况。

（1）case 语句后的各常量表达式的值不能相等，否则会出现同一个条件有多种执行方案的矛盾情况。

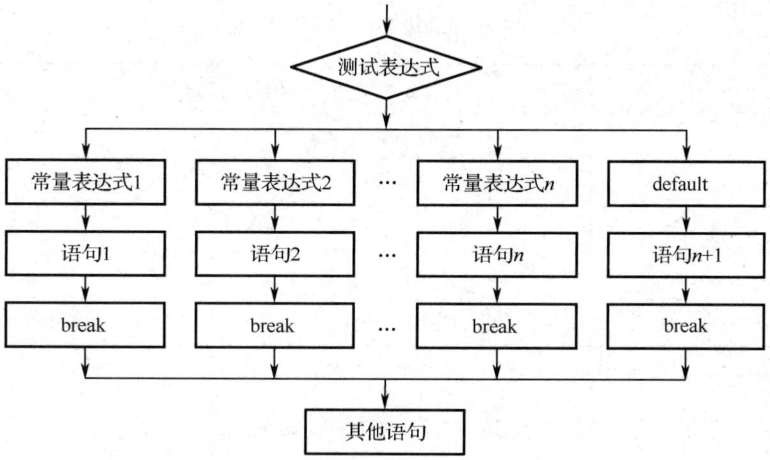

图 4-5 switch 语句的执行流程图

（2）在 case 语句后允许有多个语句，可以不用花括号括起来。例如，case 0: P1_0=1; P1_7=0; break。

（3）"case 常量表达式"只相当于一个语句标号，若测试表达式的值和某个语句标号相等，则转向该语句标号执行，但在执行完语句标号后面的语句后，不会自动跳出 switch 语句的执行，而是继续执行后面的 case 语句。因此，使用 switch 语句时，要在每个 case 语句后面加一个 break 语句，使得执行完该 case 语句后可以跳出 switch 语句的执行。

（4）case 语句和 default 语句的先后顺序可以改变，不会影响程序的执行结果。

（5）default 语句是在不满足 case 语句情况下的一个默认执行语句。如果 default 语句后面是空语句，则表示不做任何处理，可以省略。

4.1.3 八路抢答器的任务实施

【设计要求】

通过单片机设计一个八路抢答器，设计要求如下。

（1）系统上电后数码管为熄灭状态。

（2）当按下主持人按键时，数码管显示"—"，此时允许 8 位选手进行抢答，系统能够将最先抢答的选手号通过数码管显示（若 7 号选手按下选手按键，则数码管显示 7），并锁定系统禁止其他选手抢答。

（3）当再次按下主持人按键时，系统清除前一次的选手号，数码管重新显示"—"，新一轮抢答开始。

【任务分析】

根据设计要求，单片机需要实时读取 P1.0 口的状态值，从而判断是否有按键被按下，若确实有按键被按下，则将数码管的值加 1，并判断数码管的值是否有溢出，若溢出，则恢复为 0。

【实施步骤】

1. 添加元器件

打开 Proteus 仿真软件，按照表 4-1 添加元器件。

表 4-1　八路抢答器的元器件清单

元器件名称	所属类	所属子类
AT89C51	Microprocessor ICs	8051 Family
RES	Resistors	Generic
BUTTON	Switches & Relays	Switches
7SEG-MPX1-CA	Optoelectronics	7-Segment Displays
RESPACK-8	Resistors	Resistor Packs

2. 绘制仿真图

元器件全部添加后，在 Proteus ISIS 的原理图编辑窗口中按图 4-6 绘制八路抢答器仿真图。

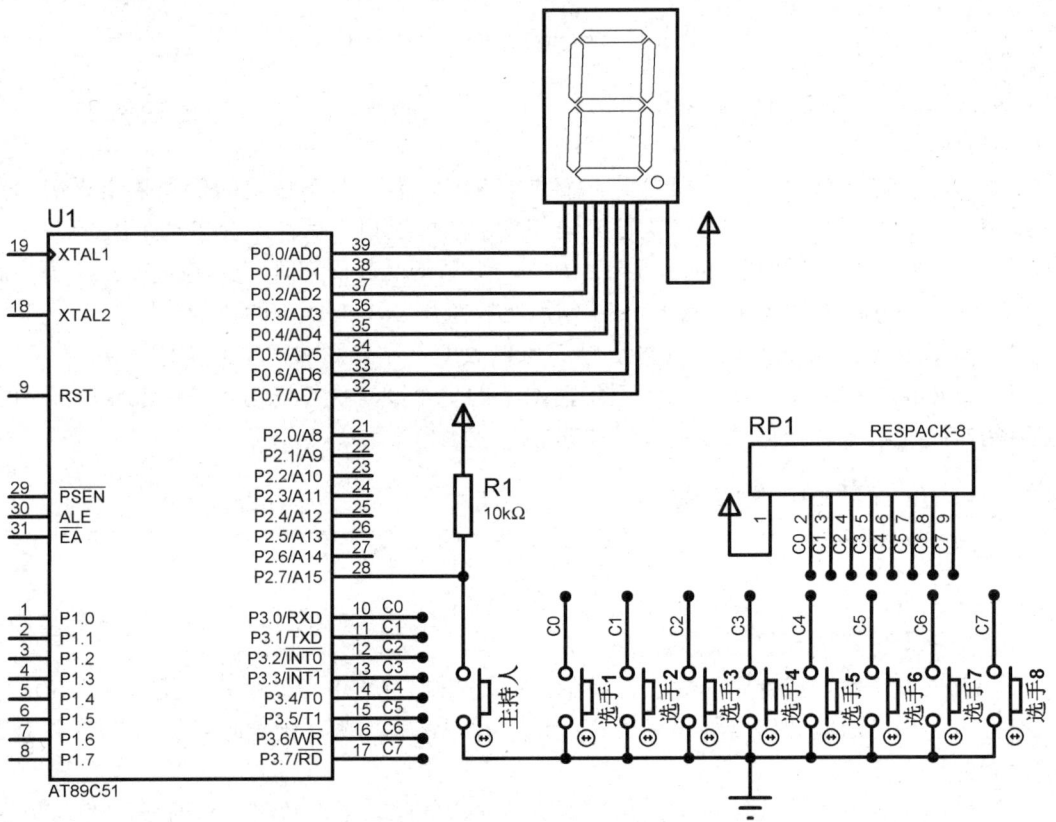

图 4-6　八路抢答器仿真图

3. 编写程序

在按键查询函数中，需要实现两种功能，一种是主持人按键的功能，另一种是 8 位选手按键的功能，主持人按键可以采用 if 语句实现，8 位选手按键可以采用 switch 语句实现，由于 8 位选手按键均接在单片机的 P3 口上，因此可以按照表 4-2，通过读取 P3 口的值判断哪个选手按键被按下。

表 4-2 P3 口所对应的 8 位选手按键的状态

序号	P3 口的状态		选手按键的状态
	二进制数	十六进制数	
1	1111 1110	0xfe	选手 1 按键被按下
2	1111 1101	0xfd	选手 2 按键被按下
3	1111 1011	0xfb	选手 3 按键被按下
4	1111 0111	0xf7	选手 4 按键被按下
5	1110 1111	0xef	选手 5 按键被按下
6	1101 1111	0xdf	选手 6 按键被按下
7	1011 1111	0xbf	选手 7 按键被按下
8	0111 1111	0x7f	选手 8 按键被按下

按键查询的流程图如图 4-7 所示。首先判断主持人按键是否被按下，若被按下，则延时 20ms 去抖动，然后再次判断主持人按键是否被按下，若被按下，则将抢答标志位置位，数码管显示"—"，接着判断抢答标志位是否有效，若无效，则直接返回，若有效，则给 P3 口赋值 0xff，然后判断 P3 口的值是否为 0xff，若是，则表示没有按键被按下，直接返回，若不是，则延时 20ms 去抖动后，再次判断 P3 口的值是否为 0xff，若是，则表示没有按键被按下，直接返回，若不是，则采用 switch 语句依次判断 P3 口的值，并选择相应的选手按键分支程序。

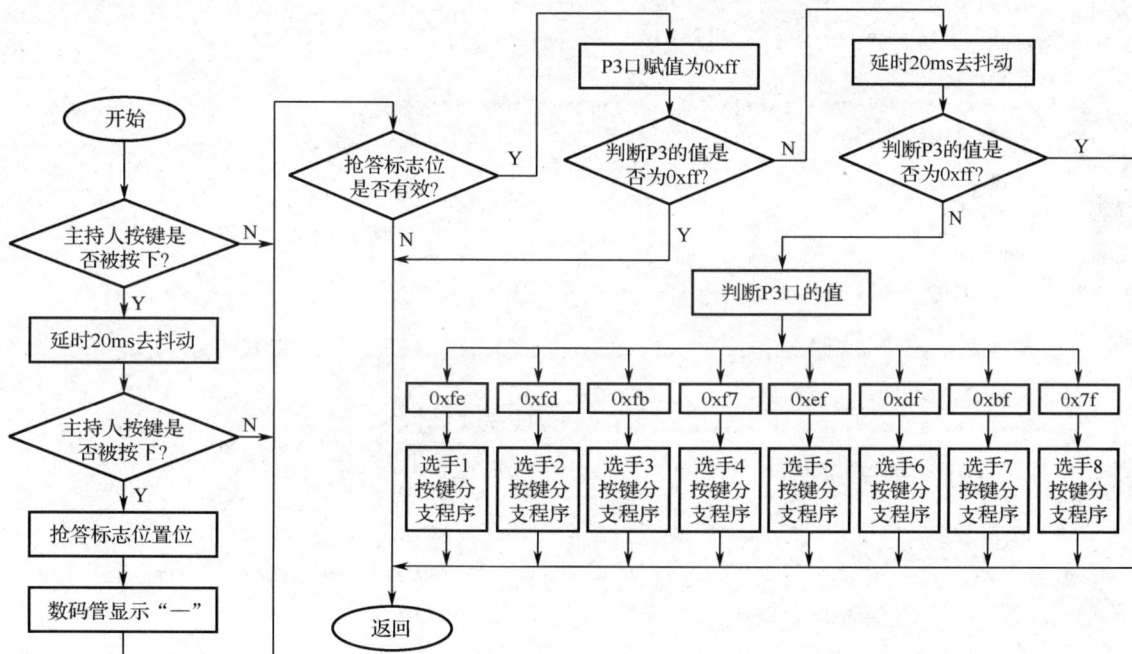

图 4-7 按键查询的流程图

在 Keil μVision5 中编写程序，实现八路抢答器的效果，参考程序如下：

```
/*******************************************************************
程序名称：program4-2.c
程序功能：八路抢答器程序
*******************************************************/
#include<reg52.h>                          //加载头文件
/*******************************************************
数据类型定义
*****************************************************/
#define uchar unsigned char                //定义无符号字符型
/*********************************************************************
变量定义
*******************************************************************/
uchar PLAYER_NUMBER=0x0b;                   //选手号，初始值处于灭的状态
bit FLAG_QIANGDA=0;                         //抢答标志位，为 0 表示禁止抢答，为 1 表示允许抢答
/*******************************************************************
数组定义：共阳极数码管 0～9、—、灭的编码表
*******************************************************************/
uchar code LED7SEG_CA[]=
{    //0    1    2    3    4    5    6    7    8    9    —    灭
    0xc0, 0xf9, 0xa4, 0xb0, 0x99, 0x92, 0x82, 0xf8, 0x80, 0x90, 0xbf, 0xff
};
/*******************************************************************
函数名称：延时函数
功能描述：延时 t ms（晶振频率为 12MHz）
入口参数：t
*******************************************************************/
void delayms(uint t)
{
    uchar i;
    while(t--)
        for(i=96;i>0;i--);                 //实现 1ms 延时
}
/*******************************************************************
选手按键查询函数
*******************************************************************/
void key_player(uchar i,j)
{
    while(P3==i);
    FLAG_QIANGDA=0;                        //将抢答标志位复位，禁止抢答
    PLAYER_NUMBER=j;                       //数码管显示选手号
}
/*******************************************************************
主持人按键查询函数
*******************************************************************/
void key_scan()
```

八路抢答器参考程序

```
{
    if(KEY_MANAGE==0)                         //判断主持人按键是否被按下
    {
        delayms(20);                          //延时 20ms 去抖动
        if(KEY_MANAGE==0)                     //再次判断主持人按键是否被按下
        {
            while(KEY_MANAGE==0);             //判断主持人按键是否被释放
            FLAG_QIANGDA=1;                   //抢答标志位置位，表示允许抢答
            PLAYER_NUMBER=0x0a;               //数码管显示"—"
        }
    }
    else if(FLAG_QIANGDA==1)                  //判断抢答标志位是否有效
    {                                         //有效则进入选手按键的判断
        P3=0xff;                              //将 P3 口设置为读入状态
        if(P3!=0xff)                          //判断选手按键是否被按下
        {
            delayms(20);                      //延时 20ms 去抖动
            if(P3!=0xff)                      //再次判断选手按键是否被按下
            {
                switch(P3)                    //查询 P3 口的状态
                {
                    case 0xfe: key_player(0xfe,1);break;   //选手 1 按键分支程序
                    case 0xfd:key_player(0xfd,2);break;    //选手 2 按键分支程序
                    case 0xfb:key_player(0xfb,3);break;    //选手 3 按键分支程序
                    case 0xf7:key_player(0xf7,4);break;    //选手 4 按键分支程序
                    case 0xef:key_player(0xef,5);break;    //选手 5 按键分支程序
                    case 0xdf:key_player(0xdf,6);break;    //选手 6 按键分支程序
                    case 0xbf:key_player(0xbf,7);break;    //选手 7 按键分支程序
                    case 0x7f:key_player(0x7f,8);break;    //选手 8 按键分支程序
                }
            }
        }
    }
}
/**********************************************************************
主函数
**********************************************************************/
void main()
{
    P0=LED_7SEG_CA[PLAYER_NUMBER];          //显示
    while(1)
    {
        key_scan();                          //查询按键
        P0=LED_7SEG_CA[PLAYER_NUMBER];       //显示
    }
}
```

4. 系统仿真

当 Keil C51 编译成功后，会自动产生 HEX 文件，接着打开之前绘制的 Proteus 仿真图，双击 AT89C51，弹出"Edit Component"对话框，单击"Program File"中的文件夹按钮，在弹出的"Select File Name"对话框中，选择之前编译生成的 HEX 文件，单击"打开"按钮，返回"Edit Component"对话框，单击"OK"按钮，即可装入 HEX 文件。

接着单击 Proteus ISIS 编辑界面左下角的运行按钮 ▶，即可观察是否能够实现八路抢答器的显示效果，如图 4-8 所示。

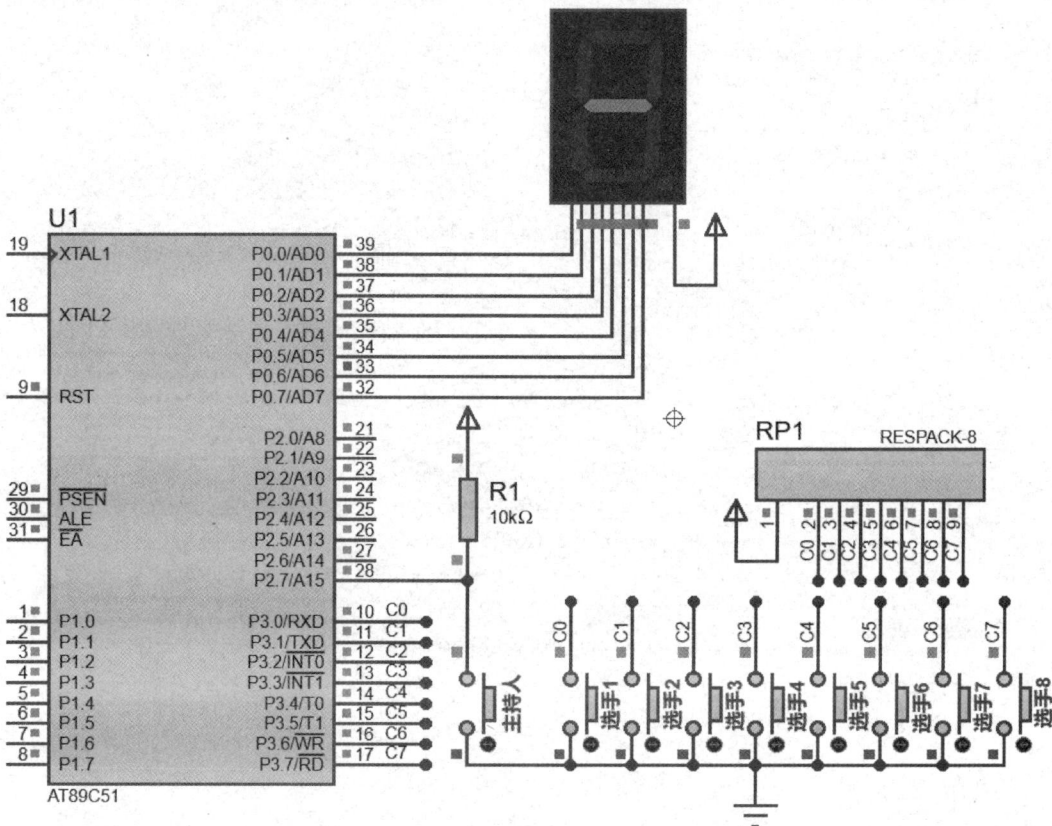

八路抢答器仿真
效果视频

图 4-8　八路抢答器仿真效果图

任务 4.2　十六路抢答器的仿真设计

✎ 学习目标

【知识目标】

（1）了解并掌握矩阵式键盘的识别原理。

（2）了解并掌握矩阵式键盘的使用方法。

十六路抢答器导学材料

【技能目标】

（1）了解并掌握单片机仿真软件 Proteus 的使用方法。

（2）了解并掌握单片机编译软件 Keil C51 的使用方法。

（3）通过十六路抢答器的仿真设计进一步掌握单片机项目的开发步骤。

【思政目标】

通过讲解矩阵式键盘，引导学生了解矩阵式键盘的工作原理、设计思路和技术挑战，从而培养他们的科学思维和探索未知领域的兴趣。

矩阵式键盘的
识别微课视频

4.2.1 矩阵式键盘的识别

独立按键电路中每个按键都占用一根 I/O 接口线。当按键数较多时，要占用较多的 I/O 接口线。因此，在按键数大于 8 时，通常采用矩阵式（也称为"行列式"）键盘电路。

1. 矩阵式键盘的工作原理

在按键数较多时，为了减少 I/O 接口线的占用，通常将按键排列成矩阵形式，如图 4-9 所示，每条水平线和垂直线在交叉处不直接相接，而是通过一个按键加以连接。这样，原本一个端口最多只有 8 个按键，现在可以构成 4×4=16 个按键，比直接将端口线用于键盘多出了一倍，而且线数越多，区别就越明显。由此可见，在需要的按键数比较多时，可采用矩阵式键盘。

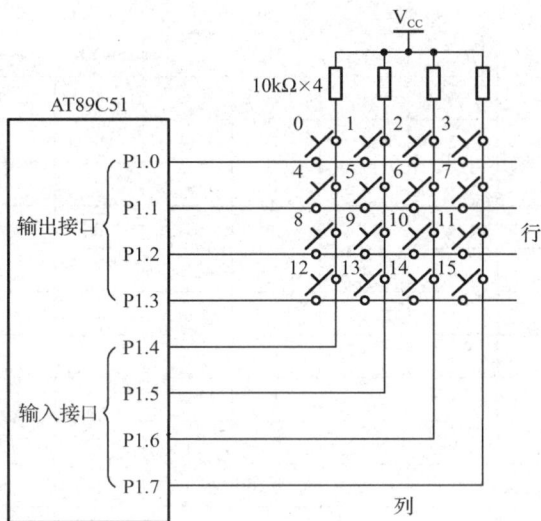

图 4-9 矩阵式键盘连接图

由图 4-8 可知，矩阵式键盘显然比独立按键要复杂一些，电路由单片机的 P1 口高、低 4 位构成 4×4 矩阵式键盘。键盘的列线一端通过电阻接正电源，另一端接单片机的输入接口线；行线一端接单片机的输出接口线，另一端悬空。故 P1.0～P1.3 作为键盘的扫描输出接口线；P1.4～P1.7 作为键盘的输入接口线。

为判断是否有按键被按下，所有的输出接口都向行线输出低电平，一旦有按键被按下，输入接口线就会被拉低，这样，通过读入输入接口线的状态就可得知是否有按键被按下了。结合图 4-9，检测的方法是将 P1.0～P1.3 输出全"0"，然后读取 P1.4～P1.7 的状态，若 P1.4～

P1.7 全为"1"，则没有按键被按下，否则肯定有按键被按下。

若有按键被按下，则判断按键的位置。当有按键被按下时，该按键处的行线和列线被接通，使穿过闭合按键的列线变为行线的状态，因此只要将 4 条行线分别设置为"0"，然后读取各列线的状态，即可判断按键的位置。具体方法是对键盘的行线进行扫描，在任意情况下，4 条行线有且只有一条被设置为"0"，其余均被设置为"1"，如表 4-3 所示。

表 4-3 矩阵式键盘各行线的输出情况

序号	P1.3	P1.2	P1.1	P1.0	说明
1	1	1	1	0	扫描第 1 行
2	1	1	0	1	扫描第 2 行
3	1	0	1	1	扫描第 3 行
4	0	1	1	1	扫描第 4 行

单片机按照表 4-3，每扫描一行，就读取各列线的值，判断各列线是否有低电平。若全为"1"，则表示这一行没有按键被按下，否则有按键被按下。由此得到闭合按键的行值和列值，然后采用计算法或查表法，将闭合按键的行值和列值转换为所定义的按键值，如表 4-4 所示。

表 4-4 4×4 矩阵式键盘的状态判断

序号	P1 口的状态									识别按键值	说明
	P1.7	P1.6	P1.5	P1.4	P1.3	P1.2	P1.1	P1.0	十六进制数		
1	1	1	1	0	1	1	1	0	0xee	按键 0	扫描第 1 行
2	1	1	0	1	1	1	1	0	0xde	按键 1	
3	1	0	1	1	1	1	1	0	0xbe	按键 2	
4	0	1	1	1	1	1	1	0	0x7e	按键 3	
5	1	1	1	0	1	1	0	1	0xed	按键 4	扫描第 2 行
6	1	1	0	1	1	1	0	1	0xdd	按键 5	
7	1	0	1	1	1	1	0	1	0xbd	按键 6	
8	0	1	1	1	1	1	0	1	0x7d	按键 7	
9	1	1	1	0	1	0	1	1	0xeb	按键 8	扫描第 3 行
10	1	1	0	1	1	0	1	1	0xdb	按键 9	
11	1	0	1	1	1	0	1	1	0xbb	按键 10	
12	0	1	1	1	1	0	1	1	0x7b	按键 11	
13	1	1	1	0	0	1	1	1	0xe7	按键 12	扫描第 4 行
14	1	1	0	1	0	1	1	1	0xd7	按键 13	
15	1	0	1	1	0	1	1	1	0xb7	按键 14	
16	0	1	1	1	0	1	1	1	0x77	按键 15	

2. 矩阵式键盘的软件设计

矩阵式键盘的扫描程序一般包含以下几项。

（1）判断是否有按键被按下。

（2）消除按键抖动。

（3）确定闭合按键的物理位置（行号、列号）。

（4）计算闭合按键的按键值。

（5）保存闭合按键的按键值，同时转去执行该闭合按键的功能。

矩阵式键盘可以采用行扫描法或列扫描法，其中行扫描法的程序流程图如图 4-10 所示。

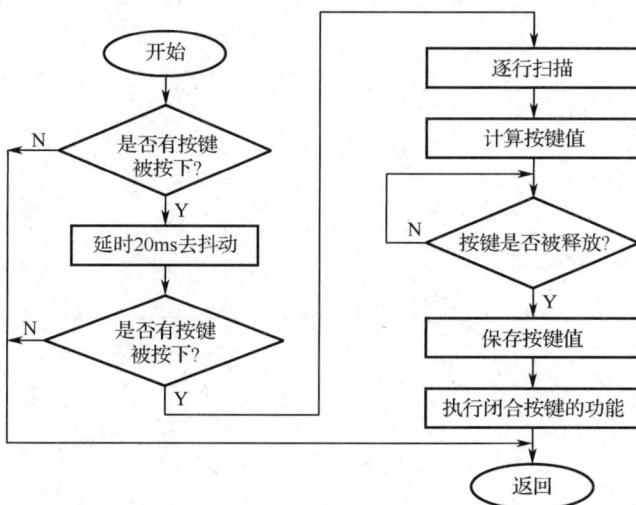

图 4-10　行扫描法的程序流程图

【课堂实训】编程实现矩阵式键盘控制数码管显示效果，单片机的 P1 口接一个 4×4 矩阵式键盘，P2、P3 口驱动 2 个独立的共阳极数码管。要求系统复位时，数码管不显示，当按动键盘时，数码管显示相应的按键号，电路连接图如图 4-11 所示。

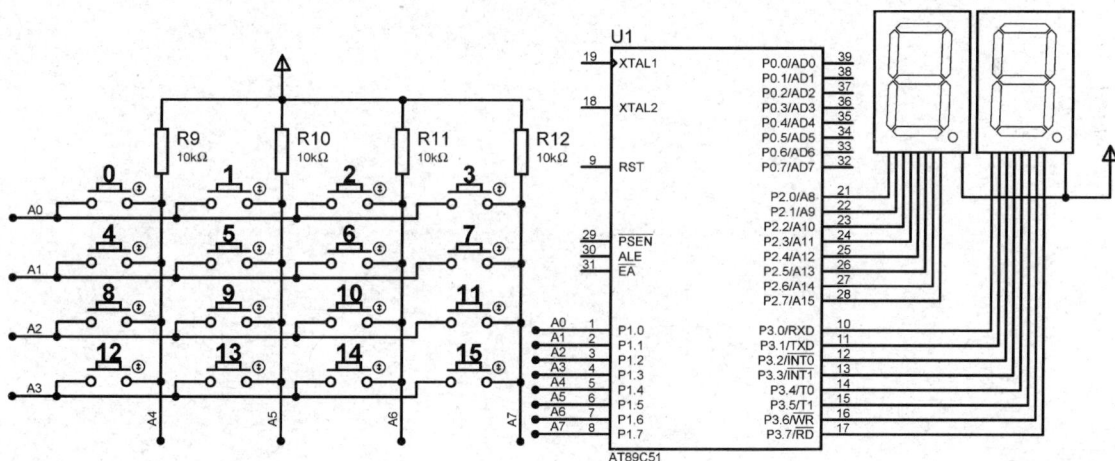

图 4-11　电路连接图

参考程序如下：

```
/**************************************************************
程序名称：program4-3.c
程序功能：矩阵式键盘控制数码管显示效果
```

课堂实训
仿真效果视频

```
************************************************************************/
#include<reg52.h>                                    //加载头文件
/************************************************************************
数据类型定义
************************************************************************/
#define uchar unsigned char                          //定义无符号字符型
/************************************************************************
数组定义：共阳极数码管 0～9、灭的编码表
************************************************************************/
uchar code LED7SEG_CA[]=
{     //0    1    2    3    4    5    6    7    8    9    灭
    0xc0, 0xf9, 0xa4, 0xb0, 0x99, 0x92, 0x82, 0xf8, 0x80, 0x90, 0xff
};
/************************************************************************
延时 t ms 函数
************************************************************************/
void delayms(uchar t)
{
    uchar i;
    while(t--)
         for(i=96;i>0;i--);
}
/************************************************************************
选手按键查询函数
************************************************************************/
void key_player(uchar i,j)
{
    while(P1==i);
    P2=LED7SEG_CA[j/10];
    P3=LED7SEG_CA[j%10];
}
/************************************************************************
按键查询函数
************************************************************************/
void key_scan()
{
    P1=0xf0;
    if(P1!=0xf0)                                      //判断是否有按键被按下
    {
        delayms(20);                                 //延时 20ms 去抖动
        if(P1!=0xf0)                                  //再次判断是否有按键被按下
        {
            P1=0xfe;                                  //查询第 1 行按键
            switch(P1)
            {
                case 0xee: key_player(0xee,0);break;          //按键 0
```

```
            case 0xde: key_player(0xde,1);break;              //按键 1
            case 0xbe: key_player(0xbe,2);break;              //按键 2
            case 0x7e: key_player(0x7e,3);break;              //按键 3
        }
        P1=0xfd;                                              //查询第 2 行按键
        switch(P1)
        {
            case 0xed: key_player(0xed,4);break;              //按键 4
            case 0xdd: key_player(0xdd,5);break;              //按键 5
            case 0xbd: key_player(0xbd,6);break;              //按键 6
            case 0x7d: key_player(0x7d,7);break;              //按键 7
        }
        P1=0xfb;                                              //查询第 3 行按键
        switch(P1)
        {
            case 0xeb: key_player(0xeb,8);break;              //按键 8
            case 0xdb: key_player(0xdb,9);break;              //按键 9
            case 0xbb: key_player(0xbb,10);break;             //按键 10
            case 0x7b: key_player(0x7b,11);break;             //按键 11
        }
        P1=0xf7;                                              //查询第 4 行按键
        switch(P1)
        {
            case 0xe7: key_player(0xe7,12);break;             //按键 12
            case 0xd7: key_player(0xd7,13);break;             //按键 13
            case 0xb7: key_player(0xb7,14);break;             //按键 14
            case 0x77: key_player(0x77,15);break;             //按键 15
        }
    }
  }
}
/********************************************************************
主函数
********************************************************************/
void main()
{
    P2=LED7SEG_CA[10];                                       //送显示
    P3=LED7SEG_CA[10];
    while(1)
        key_scan();                                          //查询按键
}
```

4.2.2　十六路抢答器的任务实施

【设计要求】

如图 4-12 所示，使用单片机设计一个十六路抢答器，系统上电后数码管处于灭的状态，当按下主持人按键时，数码管显示"一"，此时允许 16 位选手进行抢答，系统能够将最先抢答的选手号通过数码管显示（若 13 号选手按下选手按键，则数码管显示 13），并锁定系统禁止其他选手抢答,当再次按下主持人按键时，系统清除前一次的选手号，数码管重新显示"一"，新一轮抢答开始。

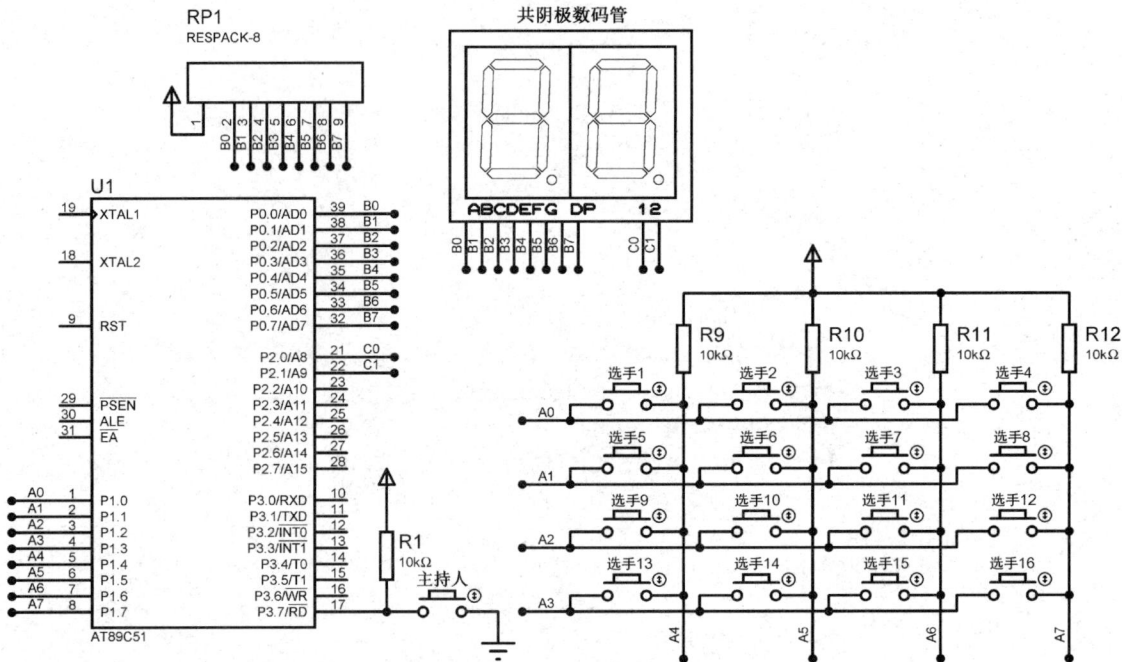

图 4-12 十六路抢答器仿真图

【任务分析】

在本任务中，主持人按键接单片机的 P3.7 口，4×4 矩阵式键盘接单片机的 P1 口，单片机的 P0 口输出二位一体共阴极数码管的段码，P2 口驱动数码管的位码端。在设计程序时，需要设置一个抢答标志位，当按下主持人按键时，抢答标志位有效，开始扫描矩阵式键盘，允许抢答，若抢答标志位无效，则不扫描矩阵式键盘，禁止抢答。

【实施步骤】

1. 添加元器件

打开 Proteus 仿真软件，按照表 4-5 添加元器件。注意：用 Proteus 仿真软件绘制单片机仿真图时，可以省略振荡电路和复位电路。

表 4-5 十六路抢答器的元器件清单

元器件名称	所属类	所属子类
AT89C51	Microprocessor ICs	8051 Family
RES	Resistors	Generic
74LS245	TTL 74LS series	Transceivers
RESPACK-8	Resistors	Resistor Packs
7SEG-MPX2-CC	Optoelectronics	7-Segment Displays
BUTTON	Switches & Relays	Switches

2. 绘制仿真图

元器件全部添加后，在 Proteus ISIS 的原理图编辑窗口中按图 4-12 绘制十六路抢答器仿真图。为了获得更好的仿真效果，在绘制仿真图时，需要将矩阵式键盘上的 4 个上拉电阻的"Model Type"设置为"DIGITAL"，如图 4-13 所示。

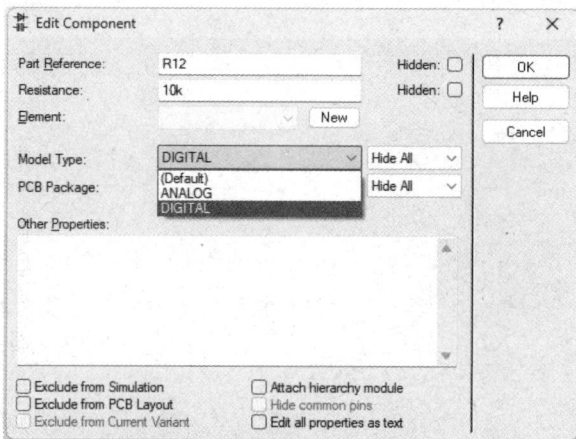

图 4-13 矩阵式键盘上拉电阻的属性设置

3. 编写程序

十六路抢答器参考程序

在 Keil μVision5 中编写程序，实现十六路抢答器的效果，参考程序如下：

```
/****************************************************************
程序名称：program4-4.c
程序功能：十六路抢答器程序
****************************************************************/
#include<reg52.h>                        //加载头文件
#include<intrins.h>                      //加载头文件
/****************************************************************
数据类型定义
****************************************************************/
#define uchar unsigned char              //定义无符号字符型
/****************************************************************
全局变量定义
****************************************************************/
uchar PLAYER_NUMBER=0xbb;                //选手号
bit FLAG_QIANGDA=0;                      //抢答标志位，为0表示禁止抢答，为1表示允许抢答
/****************************************************************
数组定义：共阴极数码管的编码表
****************************************************************/
uchar code LED_7SEG_CC[]=
{   //0    1     2     3     4     5     6     7     8     9     —     灭
    0x3f, 0x06, 0x5b, 0x4f, 0x66, 0x6d, 0x7d, 0x07, 0x7f, 0x6f, 0x40, 0x00
};
```

```
/*********************************************************************
数组定义：显示单元，2 字节
*********************************************************************/
uchar DISPLAY_DATA[2]={0,0};
/*********************************************************************
延时 t ms 函数
*********************************************************************/
void delayms(uchar t)
{
    uchar i;
    while(t--)
        for(i=96;i>0;i--);
}
/*********************************************************************
显示函数
*********************************************************************/
void display()
{
    uchar i,j=0xfe;                        //共阴驱动
    for(i=0;i<2;i++)                       //共 2 个数码管
    {
        P2=j;                             //点亮一个数码管
        P0=LED_7SEG_CC[DISPLAY_DATA[i]];  //输出数据
        j=_crol_(j,1);                    //准备点亮下一个数码管
        delayms(1);                       //延时 1ms
    }
}
/*********************************************************************
延时 20ms 去抖动
*********************************************************************/
void key_delay()
{
    uchar i;
    for(i=10;i>0;i--)                     //2ms×10=20ms
        display();                        //显示占用 2ms
}
/*********************************************************************
选手按键查询函数
*********************************************************************/
void key_player(uchar i,j)
{
    while(P1==i)
        display();
    if(FLAG_QIANGDA==1)                   //判断抢答标志位是否有效
    {                                     //若有效，则进入选手按键的处理
```

```
        FLAG_QIANGDA=0;                       //将抢答标志位复位，禁止抢答
        PLAYER_NUMBER=j;                      //数码管显示选手号
    }
}
/*************************************************************************
按键查询函数
*************************************************************************/
void key_scan()
{
    if(KEY_ZHUCHI==0)                         //判断主持人按键是否被按下
    {
        key_delay();                          //延时 20ms 去抖动
        if(KEY_ZHUCHI==0)                     //再次判断主持人按键是否被按下
        {
            while(KEY_ZHUCHI==0)              //判断主持人按键是否被释放
                display();                    //若未被释放，则处于等待状态
            FLAG_QIANGDA=1;                   //抢答标志位置位，允许抢答
            PLAYER_NUMBER=0xaa;               //数码管显示"—"
        }
    }
    P1=0xf0;
    if(FLAG_QIANGDA==1)                       //判断抢答标志位是否有效
    {
        if(P1!=0xf0)                          //判断是否有按键被按下
        {
            key_delay();                      //延时 20ms 去抖动
            if(P1!=0xf0)                      //再次判断是否有按键被按下
            {
                P1=0xfe;                      //查询第 1 行按键
                switch(P1)
                {
                    case 0xee: key_player(0xee,0x01);break;   //选手 1 按键
                    case 0xde: key_player(0xde,0x02);break;   //选手 2 按键
                    case 0xbe: key_player(0xbe,0x03);break;   //选手 3 按键
                    case 0x7e: key_player(0x7e,0x04);break;   //选手 4 按键
                }
                P1=0xfd;                      //查询第 2 行按键
                switch(P1)
                {
                    case 0xed: key_player(0xed,0x05);break;   //选手 5 按键
                    case 0xdd: key_player(0xdd,0x06);break;   //选手 6 按键
                    case 0xbd: key_player(0xbd,0x07);break;   //选手 7 按键
                    case 0x7d: key_player(0x7d,0x08);break;   //选手 8 按键
                }
                P1=0xfb;                      //查询第 3 行按键
```

```
            switch(P1)
            {
                    case 0xeb: key_player(0xeb,0x09);break;        //选手 9 按键
                    case 0xdb: key_player(0xdb,0x10);break;        //选手 10 按键
                    case 0xbb: key_player(0xbb,0x11);break;        //选手 11 按键
                    case 0x7b: key_player(0x7b,0x12);break;        //选手 12 按键
            }
            P1=0xf7;                                               //查询第 4 行按键
            switch(P1)
            {
                    case 0xe7: key_player(0xe7,0x13);break;        //选手 13 按键
                    case 0xd7: key_player(0xd7,0x14);break;        //选手 14 按键
                    case 0xb7: key_player(0xb7,0x15);break;        //选手 15 按键
                    case 0x77: key_player(0x77,0x16);break;        //选手 16 按键
                }
            }
        }
    }
}
/************************************************************************
主函数
*************************************************************************/
void main()
{
    DISPLAY_DATA[0]=PLAYER_NUMBER/16;                         //显示选手号的十位数
    DISPLAY_DATA[1]=PLAYER_NUMBER%16;                         //显示选手号的个位数
    display();                                               //显示
    while(1)
    {
            key_scan();                                      //查询按键
            DISPLAY_DATA[0]=PLAYER_NUMBER/16;                //显示选手号的十位数
            DISPLAY_DATA[1]=PLAYER_NUMBER%16;                //显示选手号的个位数
            display();                                       //显示
    }
}
```

4. 系统仿真

当 Keil C51 编译成功后，会自动产生 HEX 文件，接着打开之前绘制的
Proteus 仿真图，双击 AT89C51，弹出"Edit Component"对话框，单击"Program
File"中的文件夹按钮，在弹出的"Select File Name"对话框中，选择之前编
译生成的 HEX 文件，单击"打开"按钮，返回"Edit Component"对话框，
单击"OK"按钮，即可装入 HEX 文件。

十六路抢答器
仿真效果视频

接着单击 Proteus ISIS 编辑界面左下角的运行按钮 ▶，即可观察是否能够实现十六路
抢答器的功能。

素养小课堂

单片机的发展趋势

（1）高性能化。

处理器速度提升：随着集成电路技术的不断进步，单片机的处理器速度将持续提升，以满足更复杂的计算和控制需求。

存储容量增大：单片机的内部存储容量将不断增大，以支持更多的数据和程序存储，提升系统的整体性能。

计算能力增强：单片机将具备更强的计算能力，能够处理更加复杂的算法和数据分析任务。

（2）低功耗设计。

节能技术应用：单片机将更加注重节能技术的应用，通过优化电路设计、采用低功耗元器件等方式，降低设备的功耗，提高续航能力。

智能电源管理：引入智能电源管理技术，根据设备的实际使用情况动态地调整功耗，实现能源的高效利用。

（3）集成度提升。

功能模块集成：单片机将集成更多的功能模块，如 A/D 转换器、PWM（脉宽调制电路）、WDT（看门狗）等，以满足不同应用场景的需求，减少外围元器件的使用，降低系统成本。

SoC 趋势：随着 SoC（系统级芯片）技术的发展，单片机将逐渐向 SoC 方向发展，将整个系统的功能集成到一个芯片中，提高系统的集成度和可靠性。

（4）物联网与智能化。

物联网应用：随着物联网技术的普及和发展，单片机在智能家居、智慧城市、工业物联网等领域的应用将更加广泛和深入。单片机作为物联网设备的核心控制部件，将发挥关键作用。

智能化需求：随着经济的发展和人民生活水平的提高，人们对智能化、自动化的需求不断增加。单片机将不断适应这一趋势，为智能家居、智能穿戴、工业自动化等领域提供更加智能化的解决方案。

（5）自主创新加速。

研发步伐加快：在"自主创新"和"芯片短缺"的背景下，国内单片机厂商将加快研发步伐，逐步完成中低端领域的国产化，并持续向高端领域渗透。

市场份额提升：随着国内厂商技术实力的不断提升和市场份额的逐步扩大，国产单片机将在全球市场中占据更加重要的地位。

（6）技术创新与多样化。

技术创新：单片机行业将不断推出新技术、新产品，以满足市场不断变化的需求。例如，基于 RISC-V 等新型指令集的单片机将逐渐兴起，为行业带来新的发展机遇。

市场需求多样化：不同领域对单片机的需求各不相同，市场呈现出多样化的特点。单片机行业将针对不同领域的需求，开发具有针对性的产品，以满足市场的多样化需求。

综上所述，单片机的发展趋势将呈现高性能化、低功耗设计、集成度提升、物联网与智能化、自主创新加速及技术创新与多样化等特点，共同推动单片机行业的持续发展和进步。

课后任务

1. 编程实现一个具有倒计时功能的八路抢答器，当按下主持人按键时，抢答显示区显示"—"，30s 倒计时开始（倒计时时间为 29～0），当有选手抢答时，倒计时停止（保留倒计时数值），抢答显示区显示选手号；若在倒计时时间结束后未有选手抢答，则禁止抢答，抢答显示区显示"FF"；当再次按下主持人按键时，30s 倒计时重新开始，抢答显示区重新显示"—"。电路图如图 4-14 所示，显示效果参见二维码。

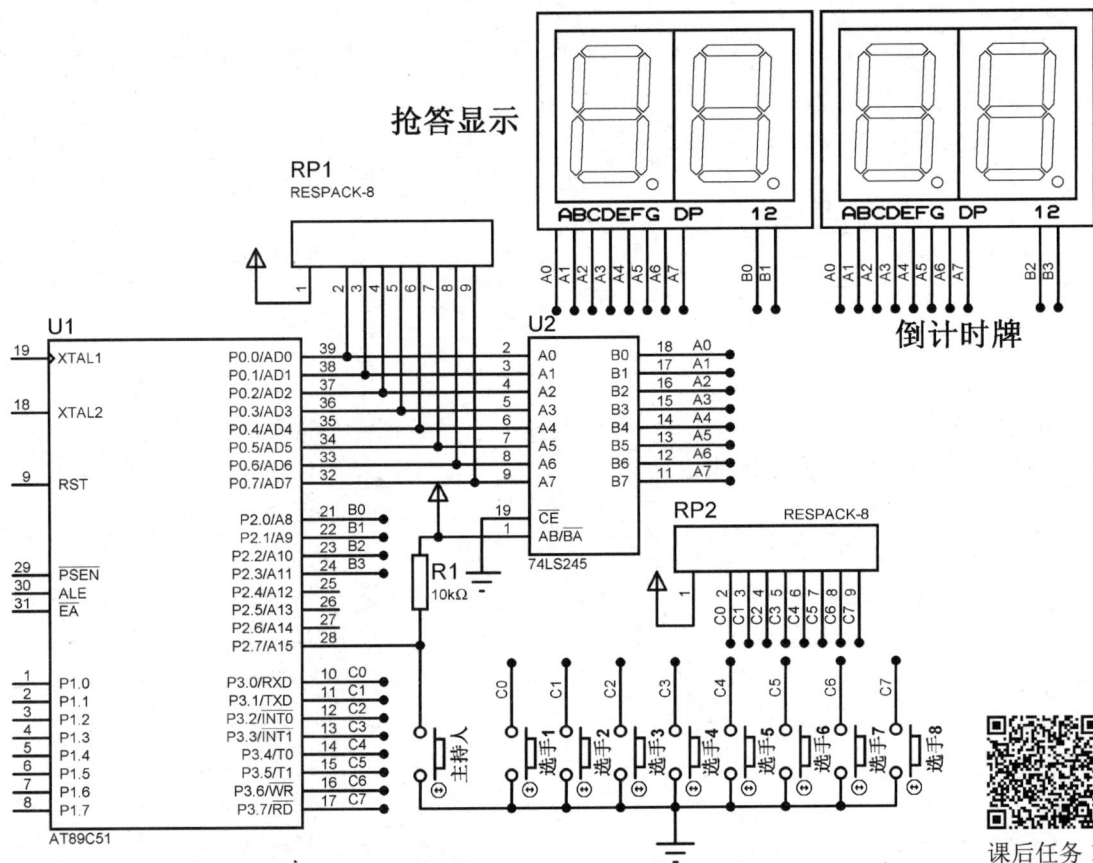

课后任务 1
仿真效果视频

图 4-14　课后任务 1 电路图

2. 编程实现一个具有倒计时功能的十五路抢答器，当按下主持人按键时，抢答显示区显示"—"，30s 倒计时开始（倒计时时间为 29～0），当有选手抢答时，倒计时停止（保留倒计时数值），抢答显示区显示选手号；若在倒计时时间结束后未有选手抢答，则禁止抢答，抢答显示区显示"FF"；当再次按下主持人按键时，30s 倒计时重新开始，抢答显示区重新显示"—"。电路图如图 4-15 所示，显示效果参见二维码。

课后任务 2
仿真效果视频

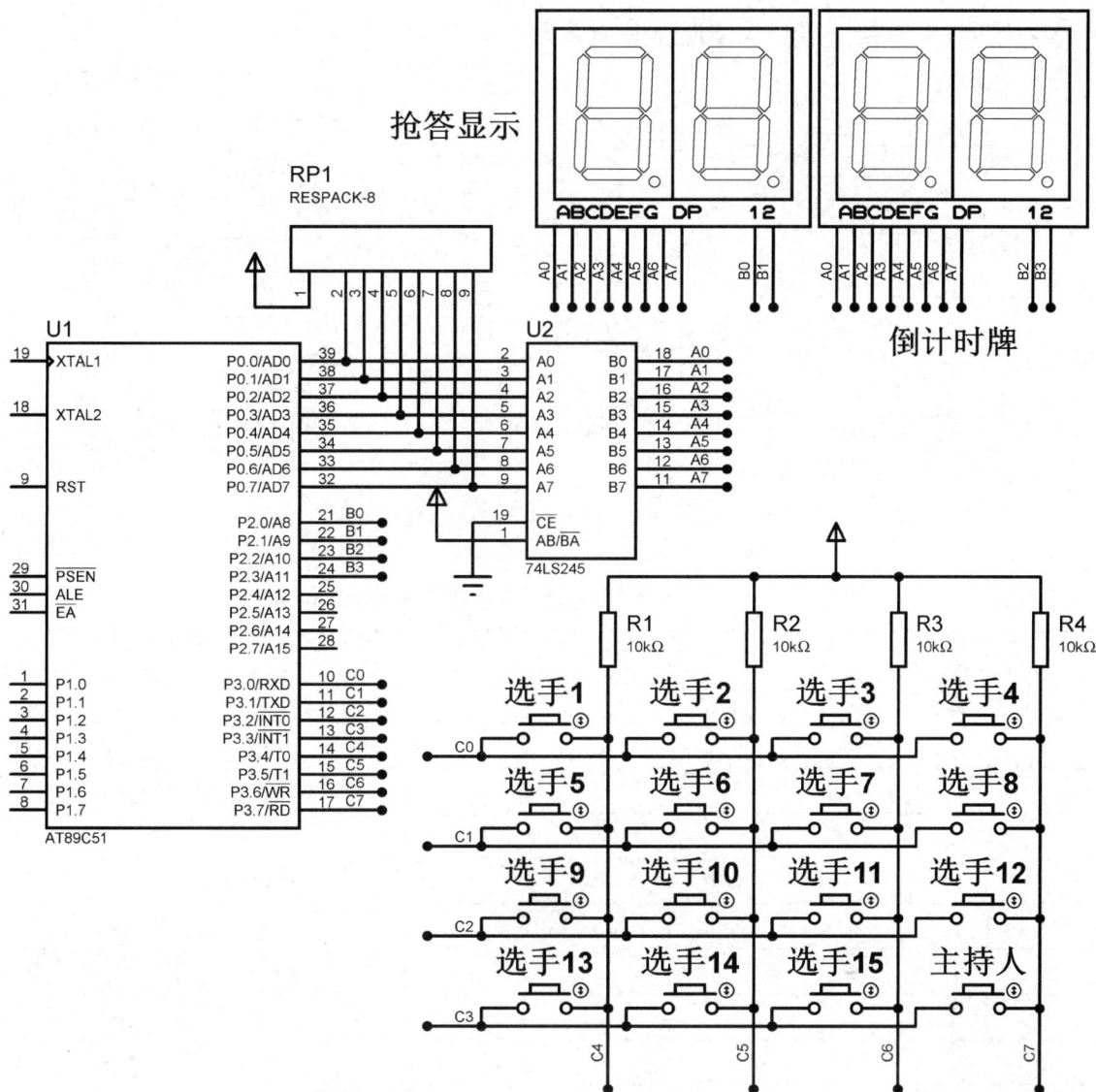

图 4-15　课后任务 2 电路图

知识拓展　4 个 I/O 接口驱动 4×4 矩阵式键盘

知识拓展
仿真效果视频

通常的 4×4 矩阵式键盘需要占用 8 个 I/O 接口，而当单片机的 I/O 接口资源比较紧张时，可以采用 4 个 I/O 接口驱动 4×4 矩阵式键盘。电路连接图如图 4-16 所示，通过单向导电的二极管，4 个 I/O 接口可以扫描 16 个按键。

4 个 I/O 接口驱动 4×4 矩阵式键盘的扫描原理如下，4 个 I/O 接口驱动 4×4 矩阵式键盘的按键值判断如表 4-6 所示。

（1）将 P1.0～P1.3 口设置为输入模式，读取这 4 个 I/O 接口的状态，若某个 I/O 接口为低电平，则表示该 I/O 接口对应的按键 3、按键 7、按键 11、按键 15 中的一个按键被按下，若这 4 个 I/O 接口均为高电平，则表示没有按键被按下。

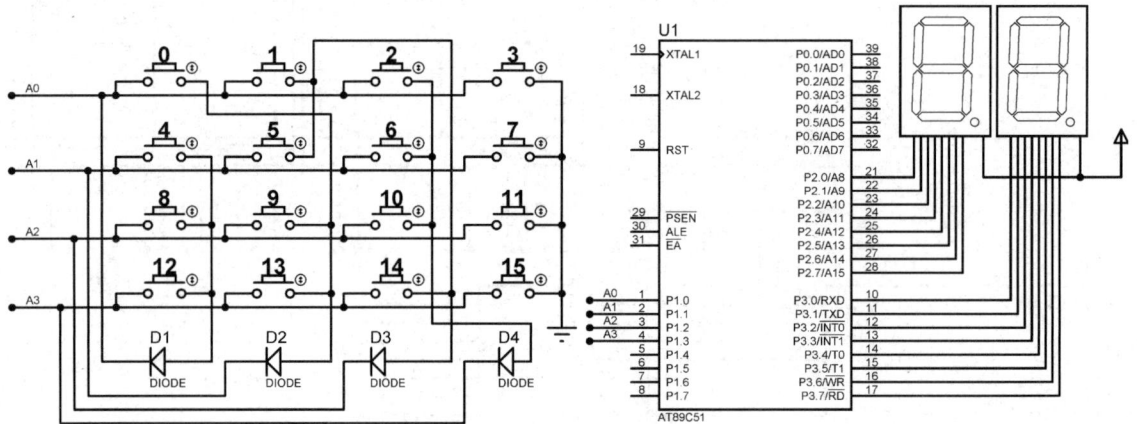

图 4-16　4 个 I/O 接口驱动 4×4 矩阵式键盘

表 4-6　4 个 I/O 接口驱动 4×4 矩阵式键盘的按键值判断

序号	P1 口的状态					识别按键值	说明
	P1.3	P1.2	P1.1	P1.0	十六进制数		
1	1	1	1	0	0xfe	按键 3	将 P1.0～P1.3 口设置为输入模式，读取这 4 个 I/O 接口的状态
2	1	1	0	1	0xfd	按键 7	
3	1	0	1	1	0xfb	按键 11	
4	0	1	1	1	0xf7	按键 15	
5	1	1	0	0	0xfc	按键 4	将 P1.0 口设置为低电平，其余设置为高电平，读取 P1.1～P1.3 口的状态
6	1	0	1	0	0xfa	按键 8	
7	0	1	1	0	0xf6	按键 12	
8	1	1	0	0	0xfc	按键 0	将 P1.1 口设置为低电平，其余设置为高电平，读取 P1.0、P1.2、P1.3 口的状态
9	1	0	0	1	0xf9	按键 9	
10	0	1	0	1	0xf5	按键 13	
11	1	0	1	0	0xfa	按键 1	将 P1.2 口设置为低电平，其余设置为高电平，读取 P1.0、P1.1、P1.3 口的状态
12	1	0	0	1	0xf9	按键 5	
13	0	0	1	1	0xf3	按键 14	
14	0	1	1	0	0xf6	按键 2	将 P1.3 口设置为低电平，其余设置为高电平，读取 P1.0～P1.2 口的状态
15	0	1	0	1	0xf5	按键 6	
16	0	0	1	1	0xf3	按键 10	

（2）将 P1.0 口设置为低电平，其余设置为高电平，读取 P1.1～P1.3 口的状态，若某个 I/O 接口为低电平，则表示该 I/O 接口对应的按键 4、按键 8、按键 12 中的一个按键被按下，若这 3 个 I/O 接口均为高电平，则表示没有按键被按下。

（3）将 P1.1 口设置为低电平，其余设置为高电平，读取 P1.0、P1.2、P1.3 口的状态，若某个 I/O 接口为低电平，则表示该 I/O 接口对应的按键 0、按键 9、按键 13 中的一个按键被按下，若这 3 个 I/O 接口均为高电平，则表示没有按键被按下。

（4）将 P1.2 口设置为低电平，其余设置为高电平，读取 P1.0、P1.1、P1.3 口的状态，若某个 I/O 接口为低电平，则表示该 I/O 接口对应的按键 1、按键 5、按键 14 中的一个按键被按下，若这 3 个 I/O 接口均为高电平，则表示没有按键被按下。

（5）将 P1.3 口设置为低电平，其余设置为高电平，读取 P1.0～P1.2 口的状态，若某个 I/O 接口为低电平，则表示该 I/O 接口对应的按键 2、按键 6、按键 10 中的一个按键被按下，若这 3 个 I/O 接口均为高电平，则表示没有按键被按下。

注意：在使用此扫描方式时，当某个 I/O 接口正在扫描时，如果有对地的按键被按下，可能导致按键误判。例如，在扫描 P1.0 口时，恰好按键 7 被按下，导致 P1.1 口为低电平，则无法判断是按键 4 还是按键 7 被按下。因此，在按键查询函数中，需要避免这种按键误判。可以采用的方法：若正在扫描 P1.0 口，则当检测到 P1.1 口为低电平时，先判断是否有对地按键被按下，若没有，则可以正确地判断是按键 4 被按下了。同理，扫描其他 3 个 I/O 接口时，判断方法一样。

习题

一、单选题

1. 当复位按键被长期按下时，一般为（　　）状态。

A. 低电平　　　　　　B. 高电平　　　　　　C. 高阻　　　　　　D. 不确定

2. 4×3 矩阵式键盘一般需要（　　）个 I/O 接口。

A. 9　　　　　　　　B. 8　　　　　　　　C. 7　　　　　　　　D. 6

二、多选题

1. 按键去抖动的方法主要有（　　）。

A. 采用 RS 触发器　　　　　　　　　　B. 采用单稳态电路

C. 采用软件延时 10～20ms　　　　　　D. 采用集成运放电路

2. 若采用行扫描法扫描 4×4 矩阵式键盘，则以下说法中正确的是（　　）。

A. 当扫描第 1 行时，其余 3 行都应该设置为高电平

B. 当第 2 行设置为低电平时，表示正在扫描第 2 行

C. 4 条列线为输入状态，可由单片机读入进行判断

D. 矩阵式键盘可以将所有行都设置为低电平，一起进行去抖动判断

三、判断题

1. switch 语句是一种分支程序结构。　　　　　　　　　　　　　　　　（　　）

2. 当按键未被按下时，单片机对应的 I/O 接口状态为低电平。　　　　　（　　）

项目五　交通灯系统的仿真设计

任务 5.1　简易交通灯系统的仿真设计

学习目标

【知识目标】

（1）了解并掌握 51 单片机的中断系统。

（2）了解并掌握中断允许控制寄存器、中断优先级控制寄存器的使用方法。

【技能目标】

（1）了解并掌握单片机仿真软件 Proteus 的使用方法。

（2）了解并掌握单片机编译软件 Keil C51 的使用方法。

（3）了解并掌握单片机程序下载的方法。

（4）了解并掌握单片机最小系统的组成。

（5）通过简易交通灯系统的仿真设计初步了解并掌握单片机项目的开发步骤。

【思政目标】

通过简易交通灯系统的仿真设计，启发学生遵守交通规则、诚信守时的职业道德和职业素养。

简易交通灯
导学材料

5.1.1　中断系统

中断是计算机中的重要技术之一，它既和硬件有关，又和软件有关。正因为中断，才使得单片机的工作更灵活，效率更高。在程序正常运行时，单片机内部或外部常会随机或定时（如定时器发出的信号）出现一些紧急事件，在多数情况下，需要 CPU 立即响应并进行处理。为了解决这一问题，在单片机中引入了中断。

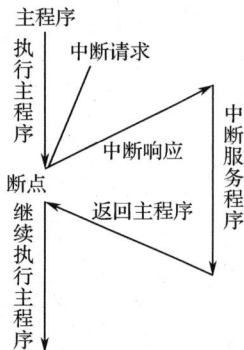

什么是中断
微课视频

图 5-1　中断的示意图

1. 中断的基本概念

（1）中断。

所谓中断是指当 CPU 正在处理某一事件 A 时，外部发生了另一事件 B，请求 CPU 迅速处理，CPU 暂时停止处理事件 A，转去处理事件 B，待 CPU 将事件 B 处理完毕后，再继续处理事件 A，这一过程被称为中断，其示意图如图 5-1 所示。

（2）主程序。

在中断系统中，通常将 CPU 在正常情况下运行的程序称为主程序或主函数。

（3）中断源。

在中断系统中，引起中断的设备或事件称为中断源。

（4）中断请求信号。

由中断源向 CPU 发出的请求中断的信号称为中断请求信号。

（5）中断响应。

CPU 接受中断请求终止现行程序而转去为服务对象服务称为中断响应。

（6）中断服务程序。

为服务对象服务的程序称为中断服务程序，也称为中断处理程序或中断函数。

（7）断点。

现行程序中断的地方称为断点。

（8）中断返回。

为服务对象服务完毕后返回原来的程序称为中断返回。

2. 引入中断的优点

单片机引入中断之后，主要具有以下优点。

（1）分时操作。

在单片机与外设交换信息时，存在高速 CPU 和低速外设（如打印机等）之间的矛盾。若采用软件查询方式，则不仅占用 CPU 操作时间，而且响应速度慢。中断解决了高速 CPU 和低速外设之间的矛盾。此时，CPU 在启动外设工作后，继续执行主程序，同时外设工作。每当外设做完一件事，就发出中断请求，请求 CPU 中断执行主程序，转去执行中断服务程序（一般是处理 I/O 数据）。中断处理完毕后，CPU 恢复执行主程序，外设仍继续工作。这样，CPU 可以命令多个外设（如键盘、打印机等）同时工作，从而提高 CPU 的工作效率。

（2）实时处理。

在实时控制中，现场的各个参数、信息是随时间和现场情况不断变化的。有了中断，外界的这些变化量就可根据要求随时向 CPU 发出中断请求，要求 CPU 及时处理，CPU 就可以马上响应（若中断响应条件满足）并加以处理。这样的及时处理在查询方式下是做不到的，大大缩短了 CPU 的等待时间。

（3）故障处理。

单片机在运行过程中，难免会出现一些无法预料的故障，如存储出错、运算溢出和电源突跳。有了中断，单片机就能自行处理，而不必停机。

3. MCS-51 系列单片机的中断源

MCS-51 系列单片机有 5 个中断源，提供两级中断优先级，可实现两级中断服务程序嵌套。对于每个中断源，根据实际需要，既可程控为高优先级中断请求或低优先级中断请求，又可程控为中断开放或中断屏蔽。中断系统的内部结构如图 5-2 所示。

从图 5-2 中可看出，MCS-51 系列单片机有 5 个中断源，即由 P3.2 和 P3.3 引脚输出的外部中断 0（$\overline{INT0}$）和外部中断 1（$\overline{INT1}$）；定时器 T0 和 T1 中断；串行接口中断（TXD 和 RXD）。下面分别进行介绍。

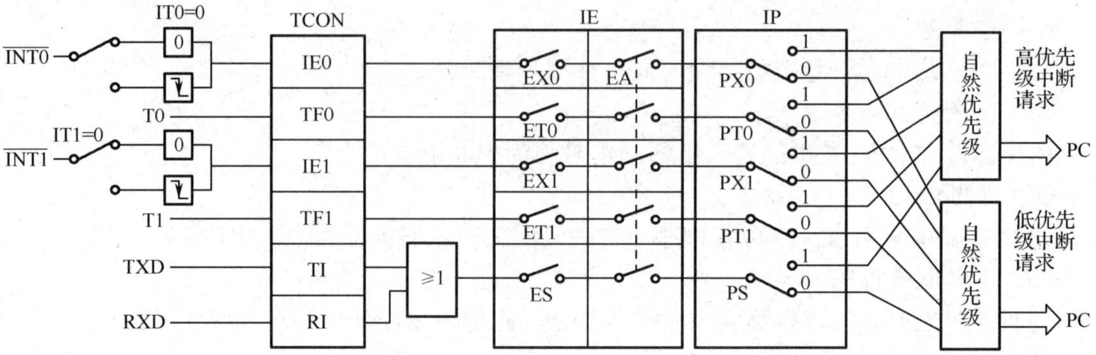

图 5-2 中断系统的内部结构

（1）外部中断。

外部中断是由外部原因引起的，包括外部中断 0 和外部中断 1，这两个中断请求信号分别通过固定引脚 $\overline{INT0}$（P3.2）和 $\overline{INT1}$（P3.3）输入。

外部中断请求信号有两种输入方式，即电平方式和脉冲方式。

在电平方式下为低电平有效，即当 P3.2 或 P3.3 引脚出现有效低电平时，外部中断标志置为 1；在脉冲方式下为下降沿有效，即当这两个引脚出现有效下降沿时，外部中断标志置为 1。注意：在脉冲方式下，外部中断请求信号的高、低电平状态都应该至少维持 1 个机器周期。

（2）定时器中断。

定时器中断是为了满足定时或计数溢出处理的需要而设置的。

定时方式的中断请求是由单片机内部引起的，输入脉冲式内部产生的周期固定的脉冲信号（1 个机器周期），无须在芯片外部设置输入端。

计数方式的中断请求是由单片机外部引起的，脉冲信号由 T0（P3.4）或 T1（P3.5）引脚输入，脉冲下降沿为计数有效信号。这种脉冲信号的周期是不固定的。

（3）串行接口中断。

串行接口中断是为了满足串行数据的传送需要而设置的。每当串行接口由 TXD（P3.1）引脚发送 1 个完整的串行帧数据，或者从 RXD（P3.0）引脚处接收 1 个完整的串行帧数据时，都会使内部串行接口中断请求标志置 1，并请求中断。

4．中断允许控制寄存器 IE

中断允许控制寄存器 IE 可以控制中断的开放和关闭。IE 的字节地址为 A8H，可以支持位寻址，各位的定义如表 5-1 所示。

表 5-1 中断允许控制寄存器 IE 各位的定义

位地址	AFH	AEH	ADH	ACH	ABH	AAH	A9H	A8H
位标志	EA	—	—	ES	ET1	EX1	ET0	EX0

（1）中断允许总控制位 EA。

当 EA=1 时，CPU 开放中断；当 EA=0 时，CPU 关闭中断。

（2）串行接口中断允许控制位 ES。

当 ES=1 时，允许串行接口中断；当 ES=0 时，禁止串行接口中断。

（3）定时器/计数器 T1 中断允许控制位 ET1。

当 ET1=1 时，允许定时器/计数器 T1 中断；当 ET1=0 时，禁止定时器/计数器 T1 中断。

（4）$\overline{INT1}$ 中断允许控制位 EX1。

当 EX1=1 时，允许 $\overline{INT1}$ 中断；当 EX1=0 时，禁止 $\overline{INT1}$ 中断。

（5）定时器/计数器 T0 中断允许控制位 ET0。

当 ET0=1 时，允许定时器/计数器 T0 中断；当 ET0=0 时，禁止定时器/计数器 T0 中断。

（6）$\overline{INT0}$ 中断允许控制位 EX0。

当 EX0=1 时，允许 $\overline{INT0}$ 中断；当 EX0=0 时，禁止 $\overline{INT0}$ 中断。

例 1： 若 51 单片机系统允许 $\overline{INT0}$、T1、串行接口中断，则 IE=？

解：如表 5-1 所示，要使得 51 单片机允许中断，则 EA=1，要允许 $\overline{INT0}$、T1、串行接口产生中断，则 EX0=1，ET1=1，ES=1，其余位为 0。

因此 IE=10011001B=99H。

5. 中断优先级控制寄存器 IP

51 单片机中断优先级的设定由中断优先级控制寄存器 IP 统一管理。它具有 2 个中断优先级，由软件设置每个中断源为高优先级中断或低优先级中断，可实现两级中断嵌套。

高优先级中断源可中断正在执行的低优先级中断服务程序，除非在执行低优先级中断服务程序时，设置了 CPU 关闭中断或禁止某些高优先级中断源的中断。同级或低优先级中断源不能中断正在执行的中断服务程序。为此，51 单片机系统内部有 2 个（用户不能访问的）优先级状态触发器，它们分别指示 CPU 是否在执行高优先级或低优先级中断服务程序，从而决定是否屏蔽所有的中断请求或同级的其他中断请求。

中断优先级控制寄存器 IP 用于选择各中断源优先级顺序，用户可用软件进行设定。其字节地址为 B8H，支持位寻址，各位的定义如表 5-2 所示。

表 5-2　中断优先级控制寄存器 IP 各位的定义

位地址	BFH	BEH	BDH	BCH	BBH	BAH	B9H	B8H
位标志	—	—	—	PS	PT1	PX1	PT0	PX0

（1）串行接口中断优先级选择位 PS。

当 PS=1 时，设定串行接口为高优先级；当 PS=0 时，设定串行接口为低优先级。

（2）T1 中断优先级选择位 PT1。

当 PT1=1 时，设定 T1 为高优先级；当 PT1=0 时，设定 T1 为低优先级。

（3）$\overline{INT1}$ 中断优先级选择位 PX1。

当 PX1=1 时，设定 $\overline{INT1}$ 为高优先级；当 PX1=0 时，设定 $\overline{INT1}$ 为低优先级。

（4）T0 中断优先级选择位 PT0。

当 PT0=1 时，设定 T0 为高优先级；当 PT0=0 时，设定 T0 为低优先级。

（5）$\overline{INT0}$ 中断优先级选择位 PX0。

当 PX0=1 时，设定 $\overline{INT0}$ 为高优先级；当 PX0=0 时，设定 $\overline{INT0}$ 为低优先级。

如果几个同级中断源同时向 CPU 申请中断，则 CPU 通过内部硬件查询逻辑按自然优先级顺序确定该响应哪个中断请求。其自然优先级由硬件形成，如表 5-3 所示。

表 5-3　各中断源及其自然优先级

编号	中断源	中断号	自然优先级
1	$\overline{INT0}$ 中断	0	最高级
2	T0 中断	1	
3	$\overline{INT1}$ 中断	2	↓
4	T1 中断	3	
5	串行接口中断	4	最低级

这种排列顺序在实际应用中很方便、合理。如果重新设置了优先级，则顺序查询逻辑电路将会相应改变排列顺序。如果给 IP 中设置的优先级控制字为 09H，则 PT1 和 PX0 均为高优先级中断源，但当这 2 个中断源同时发出中断请求时，CPU 首先响应自然优先级较高的 PX0 的中断申请。

对于中断源多于 5 个的单片机，其优先级顺序向下排列，如 AT89S52 的 T2 中断级别低于串行接口。

51 单片机的中断服务程序如下：

```
void  函数名() interrupt m (using n)
{
    中断服务内容
}
```

（1）中断服务程序是被 CPU 硬件自动调用的，而不是其他程序在代码中调用的。中断服务程序不能返回任何值，最前面用 void，后面紧跟函数名，函数名应反映其代表的功能。函数名要符合标识符的规则，可由字母、数字和下划线组成，且必须以字母或下划线开头，但不能与 C 语言中的关键字相同；中断服务程序不带任何参数，所以函数名后面的小括号内为空。

（2）关键字 interrupt 后面的 m 代表中断号，其是一个常量，取值范围是 0～4。每个中断号都对应一个中断源，如表 5-3 所示，中断号是编译器识别不同中断的唯一符号，因此在写中断服务程序时中断号务必要写正确。

（3）关键字 using 后面的 n 代表中断服务程序将选择单片机内存的哪一组工作寄存器，其也是一个常量，取值范围为 0～3。MCS-51 编译器在编译程序时会自动分配工作寄存器组，因此最后这句通常省略不写，但读者以后在遇到这样的程序代码时要知道是什么意思。

一个简单的中断服务程序写法如下：

```
void ZD_T1() interrupt 3
{
    TH1=(65536-10000)/256;
    TL1=(65536-10000)%256;
}
```

上面这段代码是定时器/计数器 T1 的一个简单的中断服务程序，定时器/计数器 T1 的中断号为 3，因此要写成 interrupt 3，中断服务程序的内容是给 2 个存放初始值的寄存器 TH1、TL1 装入新值。

5.1.2　简易交通灯系统的任务实施

【设计要求】

使用单片机设计一个简易交通灯系统,电路图如图 5-3 所示,信号变化要求如表 5-4 所示。

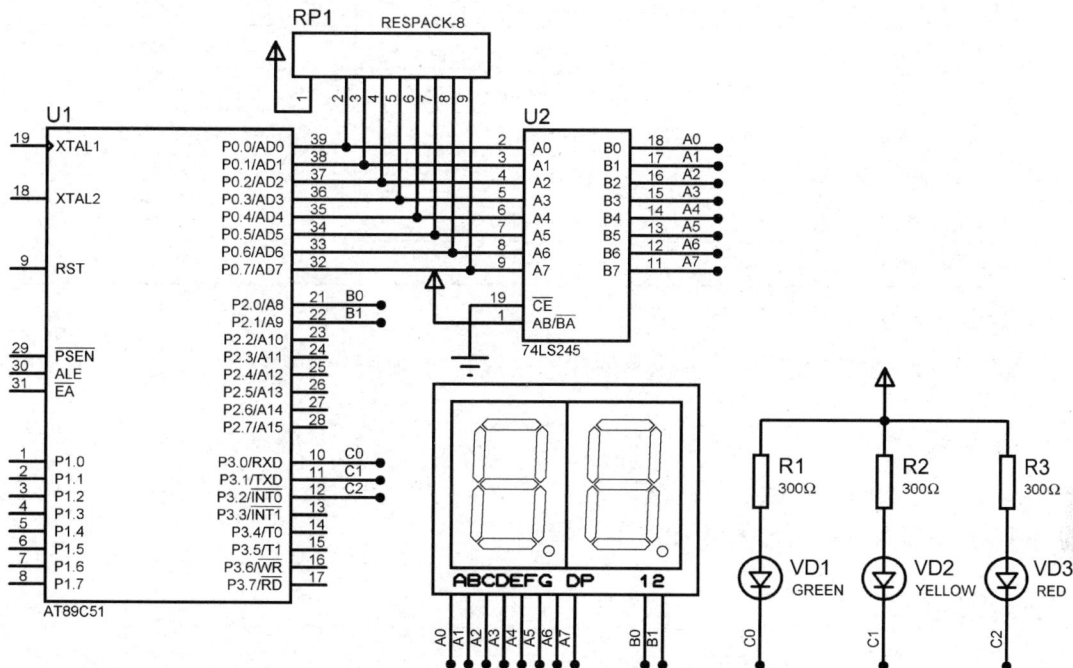

图 5-3　简易交通灯系统的电路图

表 5-4　简易交通灯系统的信号变化要求

序号	交通灯状态	倒计时时间	数码管显示范围
1	红灯亮	20s	19~0
2	黄灯亮	5s	4~0
3	绿灯亮	15s	14~0
4	黄灯亮	5s	4~0

【任务分析】

根据设计要求,采用两位一体共阴极数码管作为倒计时的显示器件,采用数码管动态扫描方式,每个数码管扫描 1ms,因此显示函数的执行时间为 2ms,将显示函数循环 500 次,即可产生 1s 的时间。

【实施步骤】

1. 添加元器件

打开 Proteus 仿真软件,按照表 5-5 添加元器件。

表5-5　简易交通灯系统的元器件清单

元器件名称	所属类	所属子类
AT89C51	Microprocessor ICs	8051 Family
RES	Resistors	Generic
7SEG-MPX2-CC	Optoelectronics	7-Segment Displays
74LS245	TTL 74LS series	Transceivers
RESPACK-8	Resistors	Resistor Packs
LED-RED	Optoelectronics	LEDs
LED-YELLOW	Optoelectronics	LEDs
LED-GREEN	Optoelectronics	LEDs

2．绘制仿真图

元器件全部添加后，在 Proteus ISIS 的原理图编辑窗口中按图 5-3 绘制简易交通灯系统仿真图。

3．编写程序

简易交通灯系统的主函数按照交通灯信号状态分为 4 个阶段：红灯亮 20s→黄灯亮 5s→绿灯亮 15s→黄灯亮 5s，其流程图如图 5-4 所示。

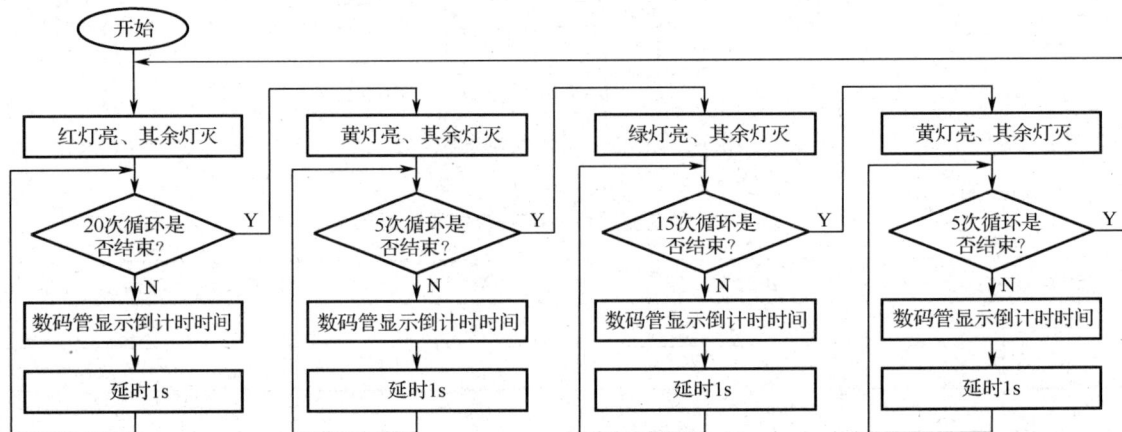

图 5-4　主函数的流程图

在 Keil μVision5 中编写程序，实现简易交通灯的效果，参考程序如下：

```
/***********************************************************
程序名称：program5-1.c
程序功能：简易交通灯系统程序
***********************************************************/
#include<reg52.h>                              //加载头文件
#include<intrins.h>                            //加载头文件
/***********************************************************
数据类型定义
```

简易交通灯系统
参考程序

```
**********************************************************************/
#define uchar unsigned char                      //定义无符号字符型
#define uint unsigned int                         //定义无符号整型
/*********************************************************************
```

数组定义：共阴极数码管的编码表

```
*********************************************************************/
uchar code LED_7SEG_CC[]=
{   //0    1     2     3     4     5     6     7     8     9
    0x3f, 0x06, 0x5b, 0x4f, 0x66, 0x6d, 0x7d, 0x07, 0x7f, 0x6f
};
/*********************************************************************
```

数组定义：显示单元，2 字节

```
*********************************************************************/
uchar DISPLAY_DATA[2]={0,0};
/*********************************************************************
```

函数名称：延时函数
功能描述：延时 t ms（晶振频率为 12MHz）
入口参数：t

```
*********************************************************************/
void delayms(uchar t)
{
    uchar i;
    while(t--)
        for(i=96;i>0;i--);                       //实现 1ms 延时
}
/*********************************************************************
```

显示函数

```
*********************************************************************/
void display()
{
    uchar i,j=0xfe;
    for(i=0;i<2;i++)
    {
        P2=j;                                     //点亮一个数码管
        P0=LED_7SEG_CC[DISPLAY_DATA[i]];          //查表输出显示数据
        delayms(1);                               //延时 1ms
        j=_crol_(j,1);                            //准备点亮下一个数码管
    }
}
/*********************************************************************
```

主函数

```
*********************************************************************/
void main()
{
    uchar i;
```

```
        uint j;
        while(1)
        {
            LED_G=1;                              //绿灯灭
            LED_Y=1;                              //黄灯灭
            LED_R=0;                              //红灯亮
            for(i=20;i>0;i--)                     //时间为20s
            {
                DISPLAY_DATA[0]=(i-1)/10;         //获得数码管的十位数
                DISPLAY_DATA[1]=(i-1)%10;         //获得数码管的个位数
                for(j=500;j>0;j--)                //2ms×500=1s
                    display();                    //调用显示函数，占用时间为2ms
            }
            LED_G=1;                              //绿灯灭
            LED_Y=0;                              //黄灯亮
            LED_R=1;                              //红灯灭
            for(i=5;i>0;i--)                      //时间为5s
            {
                DISPLAY_DATA[0]=(i-1)/10;         //获得数码管的十位数
                DISPLAY_DATA[1]=(i-1)%10;         //获得数码管的个位数
                for(j=500;j>0;j--)                //2ms×500=1s
                    display();                    //调用显示函数，占用时间为2ms
            }
            LED_G=0;                              //绿灯亮
            LED_Y=1;                              //黄灯灭
            LED_R=1;                              //红灯灭
            for(i=15;i>0;i--)                     //时间为15s
            {
                DISPLAY_DATA[0]=(i-1)/10;         //获得数码管的十位数
                DISPLAY_DATA[1]=(i-1)%10;         //获得数码管的个位数
                for(j=500;j>0;j--)                //2ms×500=1s
                    display();                    //调用显示函数，占用时间为2ms
            }
            LED_G=1;                              //绿灯灭
            LED_Y=0;                              //黄灯亮
            LED_R=1;                              //红灯灭
            for(i=5;i>0;i--)                      //时间为5s
            {
                DISPLAY_DATA[0]=(i-1)/10;         //获得数码管的十位数
                DISPLAY_DATA[1]=(i-1)%10;         //获得数码管的个位数
                for(j=500;j>0;j--)                //2ms×500=1s
                    display();                    //调用显示函数，占用时间为2ms
            }
        }
    }
```

4．系统仿真

当 Keil C51 编译成功后，会自动产生 HEX 文件，接着打开之前绘制的 Proteus 仿真图，双击 AT89C51，弹出"Edit Component"对话框，单击"Program File"中的文件夹按钮，在弹出的"Select File Name"对话框中，选择之前编译生成的 HEX 文件，单击"打开"按钮，返回"Edit Component"对话框，单击"OK"按钮，即可装入 HEX 文件。

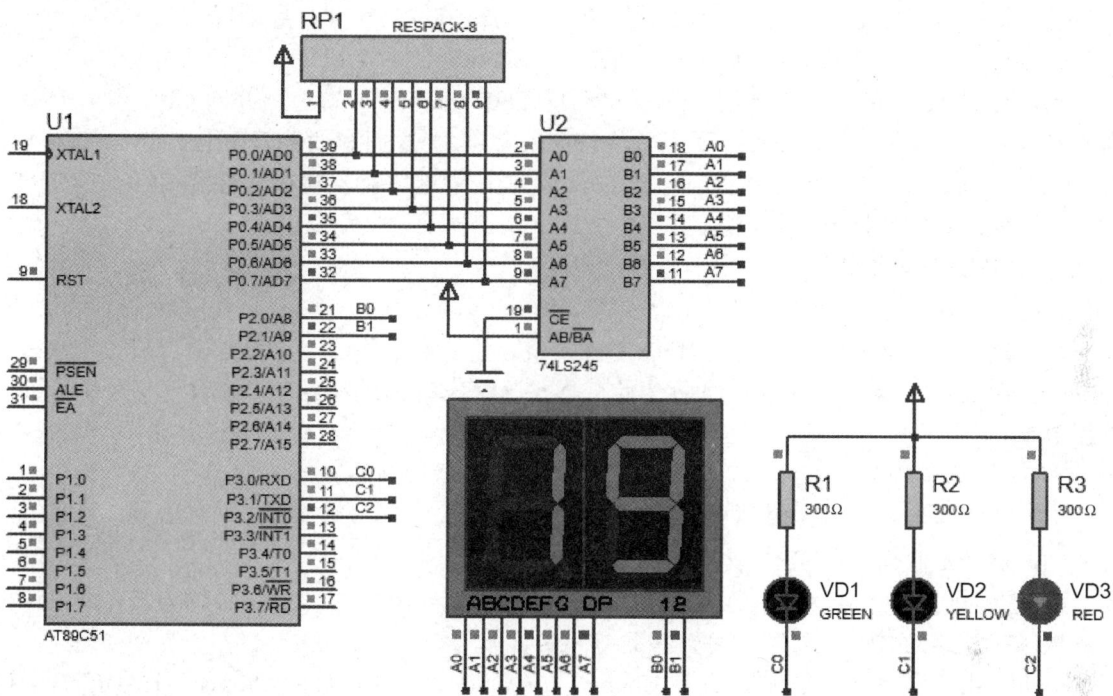

接着单击 Proteus ISIS 编辑界面左下角的运行按钮 ▶，即可观察是否能够实现简易交通灯系统的显示效果，如图 5-5 所示。

图 5-5　简易交通灯系统仿真效果图

任务 5.2　模拟交通灯系统的仿真设计

学习目标

【知识目标】

（1）了解并掌握定时器/计数器中断的原理。

（2）了解并掌握定时器/计数器中断的使用方法。

【技能目标】

（1）了解并掌握单片机仿真软件 Proteus 的使用方法。

（2）了解并掌握单片机编译软件 Keil C51 的使用方法。

（3）通过模拟交通灯系统的仿真设计进一步掌握单片机项目的开发步骤。

【思政目标】

模拟交通灯系统较为复杂，往往需要团队成员之间紧密协作，学生需要共同讨论设计方案、分工合作完成任务，并在遇到问题时相互支持、共同解决。通过模拟交通灯系统的仿真设计，培养学生的团队协作精神和沟通能力。

5.2.1 定时器中断的使用

1. 定时器中断标志位

在表 2-7 中，我们学习了 TCON 的溢出标志位和启动标志位，定时器中断也需要这两个标志位。其中，定时器 T0 的溢出标志位为 TF0，启动标志位为 TR0；定时器 T1 的溢出标志位为 TF1，启动标志位为 TR1。

定时器中断的使用
微课视频

当定时器 T0（或 T1）溢出时，硬件自动使 TF0（或 TF1）置位，并向 CPU 申请中断。当 CPU 响应中断并进入中断服务程序后，TF0（或 TF1）又被硬件自动复位。当然 TF0（或 TF1）也可用软件复位。若要启动定时器 T0（或 T1）中断，则只需用软件将 TR0（或 TR1）置位；否则，用软件将此位复位。

2. 定时器中断编程

定时器中断编程一般包含以下 4 个步骤。

（1）确定定时器的工作方式，对 TMOD 赋值。

（2）根据定时器的工作方式，通过以下公式计算定时器的初始值，其中 f_{osc} 为单片机的晶振频率：

$$定时时间 \ t=（计数器最大值 \ M-定时器初始值 \ x）\times 12/f_{osc}$$

（3）根据需要，开放定时器中断，对 IE 置初始值。

（4）启动定时器，将 TR1 或 TR0 置位。

【课堂实训】 编程实现利用定时器 T1 在 P1.0 引脚上输出周期为 100ms 的方波，设单片机的晶振频率 f_{osc}=12MHz（采用定时器中断方式实现）。

分析：由于题目要求实现周期为 100ms 的方波，因此定时时间 t=50ms，采用定时器 T1 中断方式实现，因此 TMOD=00010000B=0x10。

设定时器 T1 的初始值为 x，则

$$t = (2^{16} - x) \times \frac{12}{f_{osc}}$$

$$\Rightarrow 50 \times 10^{-3} = (2^{16} - x) \times \frac{12}{12 \times 10^{6}}$$

$$\Rightarrow x = 15536 = 0x3cb0$$

所以，TH1=0x3c，TL1=0xb0。

参考程序如下：

```
/**************************************************************
程序名称：program5-2.c
程序功能：采用定时器 T1 中断方式在 P1.0 引脚上输出周期为 100ms 的方波程序
**************************************************************/

#include<reg52.h>                          //加载头文件
```

```
/******************************************************************
引脚定义
******************************************************************/
sbit OUTPUT=P1^0;                          //单片机输出引脚定义
/******************************************************************
函数名称：T1 中断函数
功能描述：T1 中断产生周期为 100ms 的方波函数（晶振频率为 12MHz）
入口参数：无
******************************************************************/
void ZD_T1() interrupt 3                   //T1 中断函数，每 50ms 中断一次
{
    TH1=0x3c;                              //重新放入 50ms 定时器初始值
    TL1=0xb0;
    OUTPUT=~OUTPUT;                        //波形翻转
}
/******************************************************************
主函数
******************************************************************/
void main（）
{
    TMOD=0x10;                             //定时器 T1 方式 1
    TH1=0x3c;                              //放入 50ms 定时器初始值
    TL1=0xb0;
    EA=1;                                  //开启中断总开关
    ET1=1;                                 //允许 T1 中断
    TR1=1;                                 //T1 开启工作
    while(1);                              //等待中断
}
```

5.2.2 模拟交通灯系统的任务实施

【设计要求】

如图 5-6 所示，通过单片机设计一个模拟交通灯系统，模拟交通灯系统信号变化要求如表 5-6 所示。

表 5-6 模拟交通灯系统信号变化要求

序号	南北方向	东西方向	时间	南北方向数码管显示范围	东西方向数码管显示范围
1	红灯亮	绿灯亮	25s	29～5	24～0
2	红灯亮	黄灯亮	5s	4～0	4～0
3	绿灯亮	红灯亮	25s	24～0	29～5
4	黄灯亮	红灯亮	5s	4～0	4～0

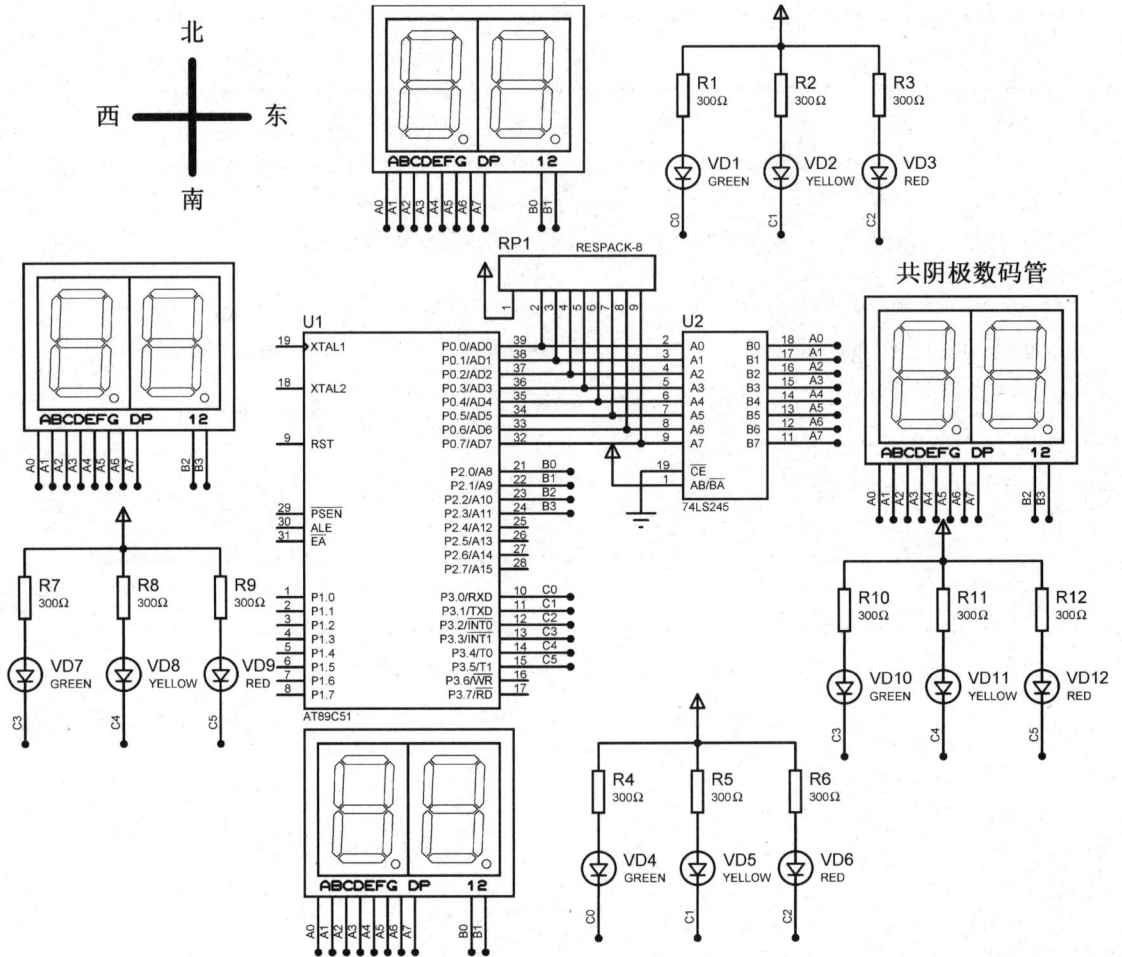

图 5-6　模拟交通灯系统电路图

【任务分析】

本任务涉及东南西北 4 个方向上的 12 盏交通灯，通过观察表 5-6 不难发现，东西方向的交通灯显示状态是一样的，同理南北方向的交通灯显示状态也是一样的。因此，可以采用 P3 口低 6 位 I/O 接口线作为模拟交通灯控制线，各控制线的分配及控制状态如表 5-7 所示。

表 5-7　模拟交通灯系统控制线的分配及控制状态

P3 口的状态							状态说明	
P3.5	P3.4	P3.3	P3.2	P3.1	P3.0	十六进制数	东西方向	南北方向
东西方向红灯	东西方向黄灯	东西方向绿灯	南北方向红灯	南北方向黄灯	南北方向绿灯			
1	1	0	0	1	1	0xf3	绿灯亮	红灯亮
1	0	1	0	1	1	0xeb	黄灯亮	红灯亮
0	1	1	1	1	0	0xde	红灯亮	绿灯亮
0	1	1	1	0	1	0xdd	红灯亮	黄灯亮

【实施步骤】

1. 添加元器件

打开 Proteus 仿真软件，按照表 5-8 添加元器件。注意：用 Proteus 仿真软件绘制单片机仿真图时，可以省略振荡电路和复位电路。

表 5-8　模拟交通灯系统的元器件清单

元器件名称	所属类	所属子类
AT89C51	Microprocessor ICs	8051 Family
RES	Resistors	Generic
74LS245	TTL 74LS series	Transceivers
RESPACK-8	Resistors	Resistor Packs
7SEG-MPX2-CC	Optoelectronics	7-Segment Displays
LED-RED	Optoelectronics	LEDs
LED-YELLOW	Optoelectronics	LEDs
LED-GREEN	Optoelectronics	LEDs

2. 绘制仿真图

元器件全部添加后，在 Proteus ISIS 的原理图编辑窗口中按图 5-6 绘制模拟交通灯系统仿真图。

模拟交通灯系统
参考程序

3. 编写程序

在 Keil μVision5 中编写程序，实现模拟交通灯系统的效果，参考程序如下：

```
/************************************************************************
程序名称：program5-3.c
程序功能：模拟交通灯系统程序
************************************************************************/
#include<reg52.h>                          //加载头文件
#include<intrins.h>                         //加载头文件
/************************************************************************
数据类型定义
************************************************************************/
#define uchar unsigned char                 //定义无符号字符型
/************************************************************************
单片机引脚定义
************************************************************************/
sbit LED_G_NB=P3^0;                         //南北方向绿灯
sbit LED_Y_NB=P3^1;                         //南北方向黄灯
sbit LED_R_NB=P3^2;                         //南北方向红灯
sbit LED_G_DX=P3^3;                         //东西方向绿灯
sbit LED_Y_DX=P3^4;                         //东西方向黄灯
sbit LED_R_DX=P3^5;                         //东西方向红灯
/************************************************************************
数组定义：共阴极数码管的编码表
************************************************************************/
```

```c
uchar code LED_7SEG_CC[]=
{   //0    1    2    3    4    5    6    7    8    9
    0x3f, 0x06, 0x5b, 0x4f, 0x66, 0x6d, 0x7d, 0x07, 0x7f, 0x6f
};
/*************************************************************************
数组定义：显示单元，4 字节
*************************************************************************/
uchar DISPLAY_DATA[4]={0, 0, 0, 0};
/*************************************************************************
延时 t ms 函数
*************************************************************************/
void delayms(uchar t)
{
    uchar i;
    while(t--)
        for(i=96;i>0;i--);
}
/*************************************************************************
显示函数
*************************************************************************/
void display()
{
    uchar i,j=0xfe;
    for(i=0;i<4;i++)
    {
        P2=j;                                 //点亮一个数码管
        P0=LED_7SEG_CC[DISPLAY_DATA[i]];      //查表输出显示数据
        delayms(1);                           //延时 1ms
        j=_crol_(j,1);                        //准备点亮下一个数码管
    }
}
/*************************************************************************
主函数
*************************************************************************/
void main()
{
    uchar i,j;
    while(1)
    {
        LED_G_NB=1;                           //南北方向绿灯灭
        LED_Y_NB=1;                           //南北方向黄灯灭
        LED_R_NB=0;                           //南北方向红灯亮
        LED_G_DX=0;                           //东西方向绿灯亮
        LED_Y_DX=1;                           //东西方向黄灯灭
        LED_R_DX=1;                           //东西方向红灯灭
        for(i=25;i>0;i--)                     //时间为 25s
        {
            DISPLAY_DATA[0]=(i+4)/10;         //获得南北方向数码管的十位数
```

```
        DISPLAY_DATA[1]=(i+4)%10;          //获得南北方向数码管的个位数
        DISPLAY_DATA[2]=(i-1)/10;          //获得东西方向数码管的十位数
        DISPLAY_DATA[3]=(i-1)%10;          //获得东西方向数码管的个位数
        for(j=250;j>0;j--)                 //4ms×250=1s
        {
            display();                     //调用显示函数，占用时间为4ms
        }
    }
    LED_G_NB=1;                            //南北方向绿灯灭
    LED_Y_NB=1;                            //南北方向黄灯灭
    LED_R_NB=0;                            //南北方向红灯亮
    LED_G_DX=1;                            //东西方向绿灯灭
    LED_Y_DX=0;                            //东西方向黄灯亮
    LED_R_DX=1;                            //东西方向红灯灭
    for(i=5;i>0;i--)                       //时间为5s
    {
        DISPLAY_DATA[0]=(i-1)/10;          //获得南北方向数码管的十位数
        DISPLAY_DATA[1]=(i-1)%10;          //获得南北方向数码管的个位数
        DISPLAY_DATA[2]=(i-1)/10;          //获得东西方向数码管的十位数
        DISPLAY_DATA[3]=(i-1)%10;          //获得东西方向数码管的个位数
        for(j=250;j>0;j--)                 //4ms×250=1s
        {
            display();                     //调用显示函数，占用时间为4ms
        }
    }
    LED_G_NB=0;                            //南北方向绿灯亮
    LED_Y_NB=1;                            //南北方向黄灯灭
    LED_R_NB=1;                            //南北方向红灯灭
    LED_G_DX=1;                            //东西方向绿灯灭
    LED_Y_DX=1;                            //东西方向黄灯灭
    LED_R_DX=0;                            //东西方向红灯亮
    for(i=25;i>0;i--)                      //时间为25s
    {
        DISPLAY_DATA[0]=(i-1)/10;          //获得南北方向数码管的十位数
        DISPLAY_DATA[1]=(i-1)%10;          //获得南北方向数码管的个位数
        DISPLAY_DATA[2]=(i+4)/10;          //获得东西方向数码管的十位数
        DISPLAY_DATA[3]=(i+4)%10;          //获得东西方向数码管的个位数
        for(j=250;j>0;j--)                 //4ms×250=1s
        {
            display();                     //调用显示函数，占用时间为4ms
        }
    }
    LED_G_NB=1;                            //南北方向绿灯灭
    LED_Y_NB=0;                            //南北方向黄灯亮
    LED_R_NB=1;                            //南北方向红灯灭
    LED_G_DX=1;                            //东西方向绿灯灭
    LED_Y_DX=1;                            //东西方向黄灯灭
    LED_R_DX=0;                            //东西方向红灯亮
```

```
    for(i=5;i>0;i--)                        //时间为 5s
    {
        DISPLAY_DATA[0]=(i-1)/10;           //获得南北方向数码管的十位数
        DISPLAY_DATA[1]=(i-1)%10;           //获得南北方向数码管的个位数
        DISPLAY_DATA[2]=(i-1)/10;           //获得东西方向数码管的十位数
        DISPLAY_DATA[3]=(i-1)%10;           //获得东西方向数码管的个位数
        for(j=250;j>0;j--)                  //4ms×250=1s
        {
            display();                      //调用显示函数，占用时间为 4ms
        }
    }
}
```

4. 系统仿真

当 Keil C51 编译成功后，会自动产生 HEX 文件，接着打开之前绘制的 Proteus 仿真图，双击 AT89C51，弹出"Edit Component"对话框，单击"Program File"中的文件夹按钮，在弹出的"Select File Name"对话框中，选择之前编译生成的 HEX 文件，单击"打开"按钮，返回"Edit Component"对话框，单击"OK"按钮，即可装入 HEX 文件。

模拟交通灯系统
仿真效果视频

接着单击 Proteus ISIS 编辑界面左下角的运行按钮 ▶，即可观察是否能够实现模拟交通灯系统的功能，如图 5-7 所示。

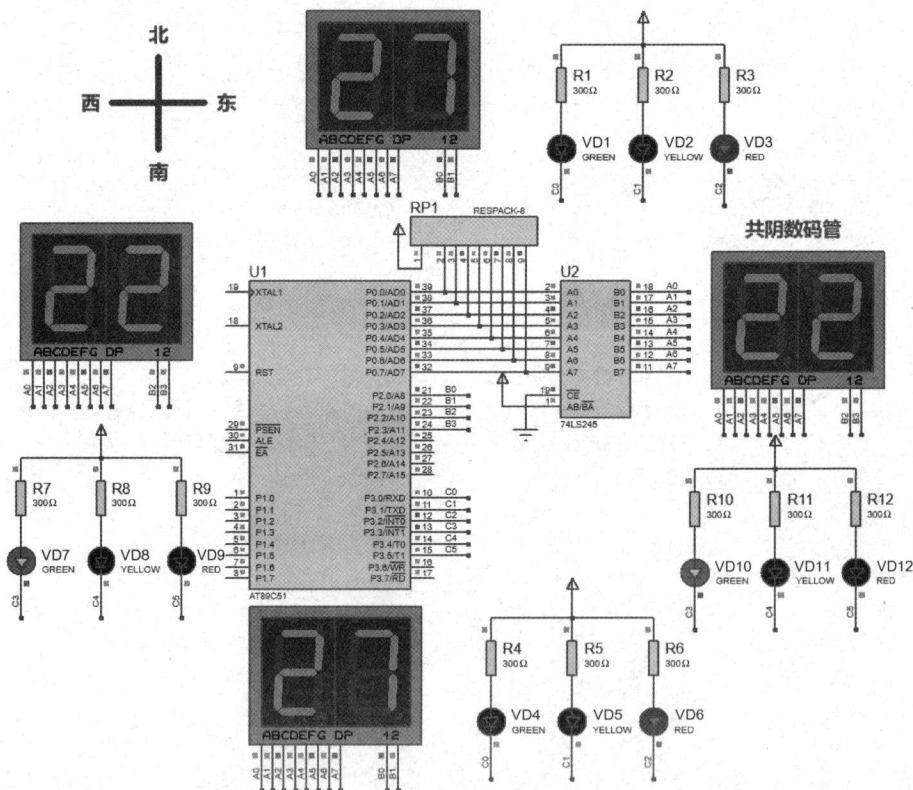

图 5-7　模拟交通灯系统仿真效果图

素养小课堂

"中国芯"发展纪要

（1）1965年，中国自主研制的第一块集成电路在上海诞生，从此中国进入集成电路时代。

（2）1970年，中国人民解放军军事工程学院成功研制出441B-III，是我国第一台具有分时操作系统和汇编语言、FORTRAN语言及标准程序库的计算机。

（3）1972年，中国自主研制的大规模集成电路在四川永川半导体研究所诞生，实现了从中小规模集成电路到大规模集成电路的跨越（美国是从1960年到1968年，用了8年时间。中国是从1965年到1972年，用了7年时间）。

（4）1978—1981年，中国科学院半导体研究所成功研制出4KB、16KB的DRAM。

（5）1986年，无锡742厂试制成功第一片64KB的DRAM，集成度为15万个元器件，加工工艺为3μm（相比日本NEC晚6年）。

（6）1999年12月，北京大学微处理器研究开发中心成功研制出国内第一套支持微处理器正向设计的开发平台，并成功研制出16位微处理器原型系统。

（7）2004年10月18日，华为旗下的海思半导体宣告成立。

（8）2020年，华为麒麟芯片产品已更新至5G高端机型上的麒麟990，而海思也迎来了鼎盛巅峰，形成了麒麟系列、鲲鹏系列、昇腾系列，以及基带产品天罡与巴龙等专用芯片。

课后任务

1. 在任务5.1的基础上添加4个按键，通过这4个按键可以调节简易交通灯系统的红灯时间（调节范围为10～30s），电路图如图5-8所示。当按下设置按键后，系统处于暂停状态，此时红灯亮，表示可以通过加按键、减按键调节红灯的倒计时时间，当按下确定按键后，表示红灯倒计时结束，系统按照最新的倒计时时间开始工作，显示效果参见二维码。

课后任务1仿真效果视频

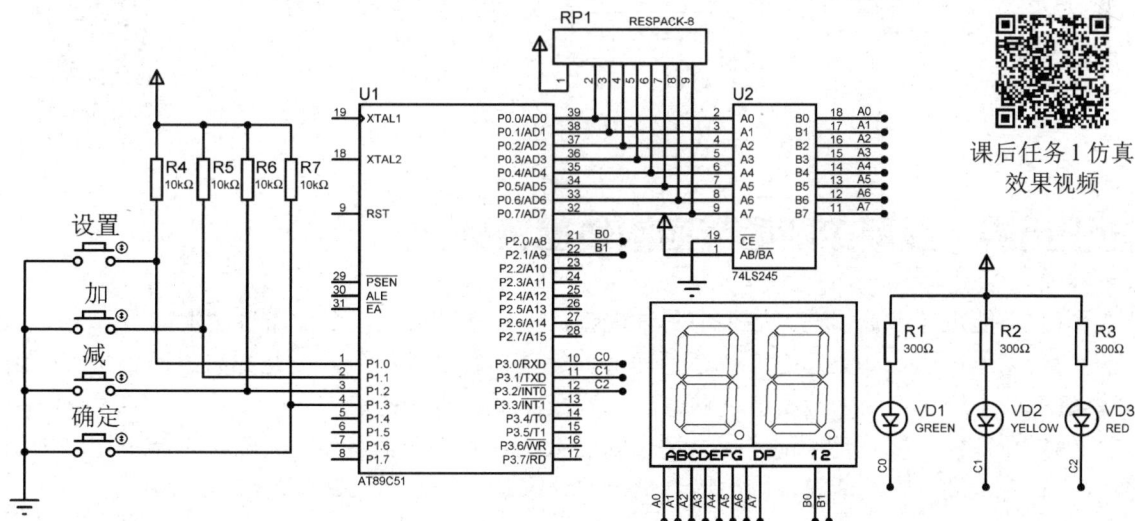

图5-8　课后任务1电路图

2．在任务 5.2 的基础上添加 4 个按键，通过这 4 个按键可以调节模拟交通灯系统的绿灯时间（调节范围为 20～40s，无论如何调节，红灯时间都比绿灯时间长 5s，黄灯时间一律为 5s），电路图如图 5-9 所示。当按下设置按键后，系统处于暂停状态，此时红灯亮，表示可以通过加按键、减按键调节绿灯的倒计时时间，当按下确定按键后，表示绿灯倒计时结束，系统按照最新的倒计时时间开始工作，显示效果参见二维码。

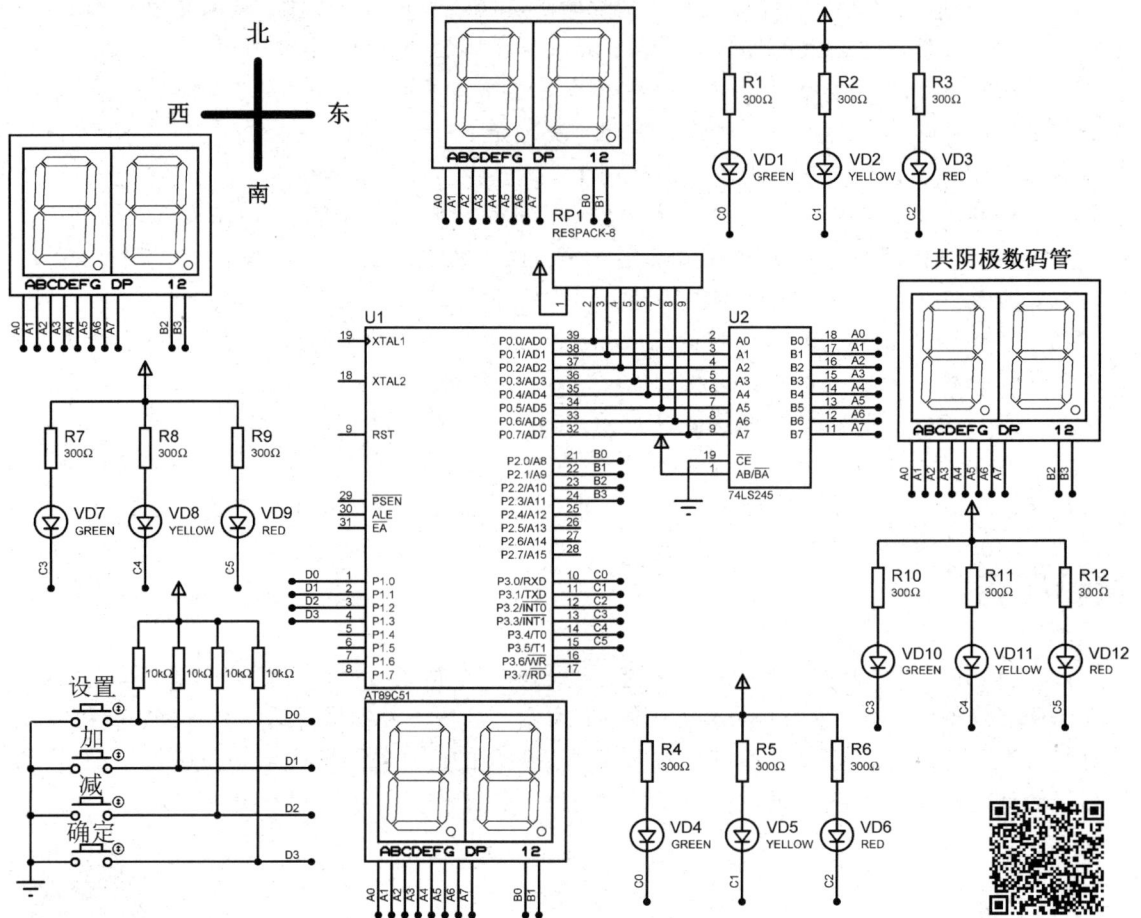

图 5-9　课后任务 2 电路图

课后任务 2 仿真
效果视频

知识拓展　编写中断函数应遵循的规则

（1）不能进行参数传递。如果中断过程包括任何参数声明，编译器将产生错误。

（2）无返回值。如果想定义一个返回值，编译器将产生错误，但是如果返回整型值，编译器将不产生错误，因为整型值是默认值，编译器不能清楚识别。

（3）在任何情况下都不能直接调用中断函数，否则编译器将产生错误。直接调用中断函数时，硬件上没有中断请求，因而这个指令的结果是不确定的，并且通常是致命的。

（4）可以在中断函数定义中使用 using 指令，用来指定当前使用的工作寄存器组。51 单片机共有 4 组工作寄存器组，程序具体使用哪一组工作寄存器由程序状态字寄存器 PSW 中的 RS1 和 RS0 来确定。不同的中断函数使用不同的工作寄存器组，可以避免中断嵌套调用时的资源冲突。

（5）在中断函数中，调用的函数所使用的工作寄存器组必须与中断函数相同，当没有使用 using 指令时，编译器会选择一个工作寄存器组作为绝对寄存器访问，程序员必须保证按要求使用相同的工作寄存器组，编译器不会对此进行检查。

习题

一、单选题

1．当 CPU 响应单片机定时器中断后，定时器 T0 的溢出标志位 TF0（　　）。

A．由硬件复位　　　　　　　　　　　　B．由软件复位

C．由硬件和软件都可以复位　　　　　　D．不能复位

2．在下列中断优先级顺序排列中，有可能实现的是（　　）。

A．T1、T0、$\overline{INT0}$、$\overline{INT1}$、串行接口

B．$\overline{INT0}$、T0、$\overline{INT1}$、T1、串行接口

C．$\overline{INT1}$、$\overline{INT0}$、串行接口、T0、T1

D．$\overline{INT1}$、串行接口、T0、$\overline{INT0}$、T1

3．下列说法中错误的是（　　）。

A．同级的中断请求按时间的先后顺序响应

B．同一时间同级的多中断请求将形成阻塞，系统无法响应

C．低优先级中断请求不能中断高优先级中断请求，但是高优先级中断请求能中断低优先级中断请求

D．同级中断不能嵌套

4．当优先级的设置相同时，若以下几个中断同时发生，则（　　）中断优先响应。

A．$\overline{INT1}$　　　　　B．T1　　　　　C．串行接口　　　　　D．T0

5．要使 MCS-51 系列单片机能够响应定时器 T1 中断、串行接口中断，IE 的字节地址应是（　　）。

A．98H　　　　　　　B．84H　　　　　　　C．42H　　　　　　　D．22H

二、多选题

1．下列关于 C51 中断函数定义格式的描述中，（　　）是正确的。

A．n 是与中断源对应的中断号，取值范围为 0～4

B．m 是工作寄存器组的组号，在默认情况下由 PSW 的 RS0 和 RS1 确定

C．interrupt 是 C51 的关键字，不能作为变量名

D．using 也是 C51 的关键字，不能省略

2．下列关于中断控制寄存器的描述中，（　　）是正确的。

A．80C51 单片机有 4 个与中断有关的控制寄存器

B．TCON 为串行接口控制寄存器，字节地址为 98H，可位寻址

C．IP 为中断优先级控制寄存器，字节地址为 B8H，可位寻址

D．IE 为中断允许控制寄存器，字节地址为 A8H，可位寻址

三、判断题

1．定时器 T0 中断可以被 $\overline{\text{INT0}}$ 中断。 （   ）

2．89C51 单片机的定时器/计数器是否工作可以通过外部中断进行控制。 （   ）

项目六　计数系统的仿真设计

任务 6.1　流水线计数系统的仿真设计

学习目标

【知识目标】

（1）了解并掌握 51 单片机计数器的原理。

（2）了解并掌握 51 单片机计数器的使用方法。

【技能目标】

（1）了解并掌握单片机仿真软件 Proteus 的使用方法。

（2）了解并掌握单片机编译软件 Keil C51 的使用方法。

（3）了解并掌握单片机程序下载的方法。

（4）了解并掌握单片机最小系统的组成。

（5）通过流水线计数系统的仿真设计初步了解并掌握单片机项目的开发步骤。

【思政目标】

通过流水线计数系统的仿真设计，在教授如何使用 51 单片机计数器进行编程时，强调一丝不苟的工匠精神，鼓励学生追求程序的精确性和效率，培养其在学习和工作中的严谨态度。

6.1.1　计数器的使用

1．51 单片机计数器

51 单片机内部有 2 个 16 位定时器/计数器，即 T0 和 T1。它们都具有定时和计数功能，可用于定时或延时控制，对外部事件进行检测、计数等。

定时器用作定时时，对机器周期进行计数，每过一个机器周期，计数器就加 1，直到计数器计满溢出。51 单片机采用同步控制方式，因此它有固定的机器周期。规定一个机器周期的宽度为 6 个状态，依次表示为 S1～S6。由于 1 个状态又包括 2 个拍节，因此一个机器周期共有 12 个拍节，分别记作 S1P1、S1P2、…、S6P2。由于一个机器周期由 12 个时钟周期组成，因此其计数频率为时钟周期的 1/12。显然，定时器的定时时间不仅与计数器的初始值即计数长度有关，还与系统的时钟频率有关。

定时器用作计数时，对来自输入引脚 T0（P3.4）和 T1（P3.5）的外部信号进行计数，在每个机器周期的 S5P2 期间采样引脚输入电平。若前一个机器周期的采样值为 1，后一个机器周期的采样值为 0，则计数器加 1。新的计数值是在检测到输入引脚电平发生 1 到 0 的负跳变后，于下一个机器周期的 S3P1 期间装入计数器的。由于它需要两个机器周期（24 个时钟周期）来识别一个 1 到 0 的跳变信号，因此最高的计数频率为时钟频率的 1/24。对外部输入信

号的占空比没有特别的限制，但必须保证输入信号电平在发生跳变前至少被采样一次，因此输入信号电平至少在一个完整的机器周期中保持不变。

TMOD 的格式如下所示，我们只要将 TMOD 的功能选择位 C/$\overline{T}$ 置为高电平，就可以将定时器/计数器置为计数功能。

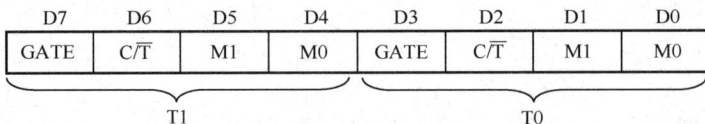

D7	D6	D5	D4	D3	D2	D1	D0
GATE	C/$\overline{T}$	M1	M0	GATE	C/$\overline{T}$	M1	M0

T1 （D7~D4）　　T0 （D3~D0）

例 1：若采用定时器 T1 方式 1 对外部脉冲计数，要求计满 1000 产生溢出，则 TMOD、TH1、TL1 分别为多少？

解：由于 T1 工作，所以 TMOD 的低 4 位均为 0，而 T1 的 GATE=0，C/$\overline{T}$=1，M1=0，M0=1。

所以，TMOD=01010000B=50H。

设 T1 的计数初始值为 x，n 为计数值，则

$$n = 2^{16} - x$$
$$\Rightarrow 1000 = 2^{16} - x$$
$$\Rightarrow x = 64536 = 0xfc18$$

所以，TH1=0xfc，TL1=0x18。

2. 计数器编程

计数器编程一般包含以下 4 个步骤。

（1）确定计数器的工作方式，对 TMOD 赋值。

（2）根据计数器的工作方式，通过以下公式计算计数器的计数初始值：

计数初始值 x=计数器最大值 M−计数值 n

（3）根据需要，开放定时器/计数器中断，对 IE 置初始值。

（4）启动定时器/计数器，将 TR1 或 TR0 置位。

【课堂实训】编程实现利用 T0 对外部脉冲进行计数，每计满 100 个周期，发光二极管亮或灭一次，采用中断方式实现，仿真图如图 6-1 所示。

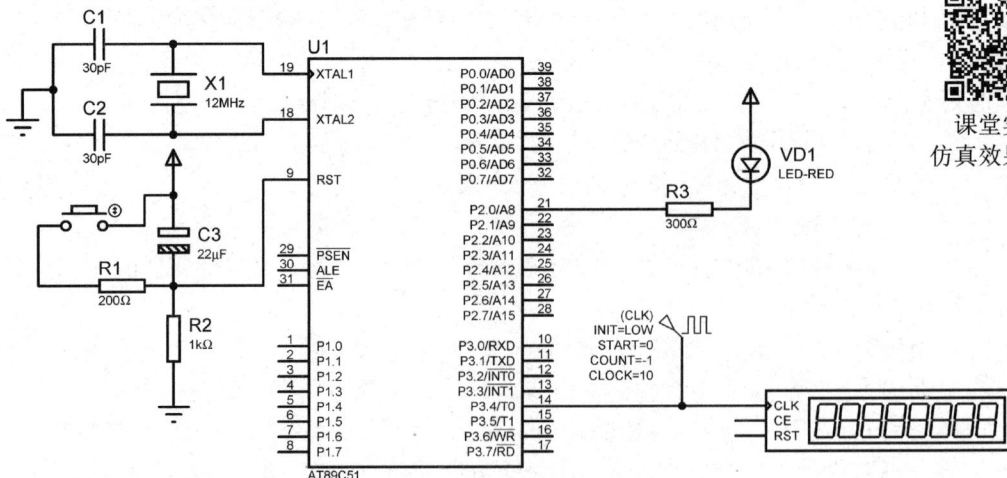

课堂实训
仿真效果视频

图 6-1　T0 对外部脉冲计数仿真图

注意：在绘制仿真图时，方波信号源"DCLOCK"的频率设置为 10Hz，其属性设置如图 6-2 所示；虚拟仪表"COUNTER TIMER"需要选择计数功能，其属性设置如图 6-3 所示。

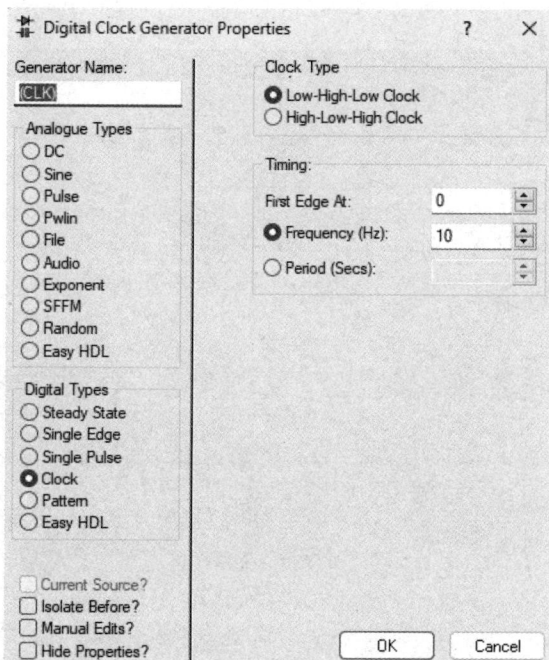

图 6-2　方波信号源"DCLOCK"的属性设置

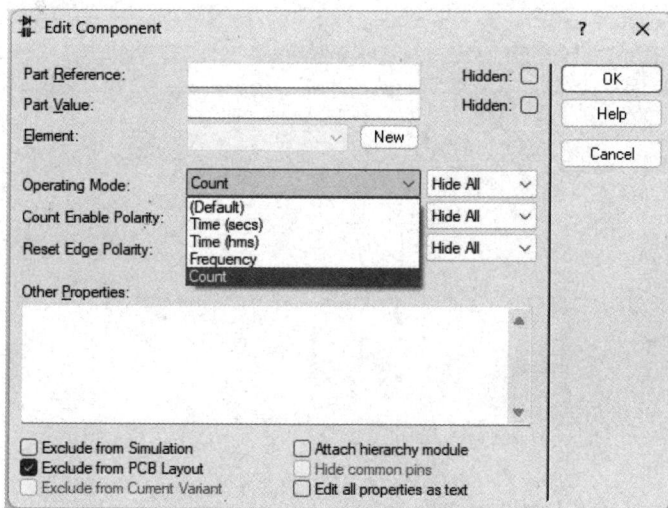

图 6-3　虚拟仪表"COUNTER TIMER"的属性设置

分析：由于题目要求对外部脉冲计满 100 个周期，因此可以采用 T0 的方式 2 来实现，TMOD=00000110B=0x06。

设 T0 的计数初始值为 x，则

$$n = 2^8 - x$$
$$\Rightarrow 100 = 256 - x$$
$$\Rightarrow x = 156$$

所以，TH0=TL0=156。

参考程序如下：

```
/***********************************************************************
程序名称：program6-1.c
程序功能：T0 对外部脉冲计数程序
***********************************************************************/
#include<reg52.h>                          //加载头文件
/***********************************************************************
引脚定义
***********************************************************************/
sbit LED=P2^0;                             //引脚定义
/***********************************************************************
函数名称：T0 中断函数
功能描述：T0 对外部脉冲计数函数
入口参数：无
***********************************************************************/
void ZD_T0() interrupt 1                   //T0 中断服务程序
{
    LED=~LED;                              //发光二极管亮或灭一次
}
/***********************************************************************
主函数
***********************************************************************/
void main()
{
TMOD=0x06;                                 //T0 的方式 2
    TH0=156;                               //T0 计满 100 个周期的初始值
    TL0=156;
    EA=1;                                  //开启中断总开关
    ET0=1;                                 //允许 T0 中断
    TR0=1;                                 //T0 开启工作
    while(1);                              //等待中断
}
```

6.1.2　流水线计数系统的任务实施

【设计要求】

使用单片机设计一个流水线计数系统，其电路图如图 6-4 所示，采用计数器 T1 对流水线上的物品进行计数，共阴极数码管实时显示流水线上的物品计数值，最大值为 30。当计数值未超过 30 时，绿灯亮、红灯灭；当计数值超过 30 时，绿灯灭、红灯亮，数码管一直显示计数最大值 30。

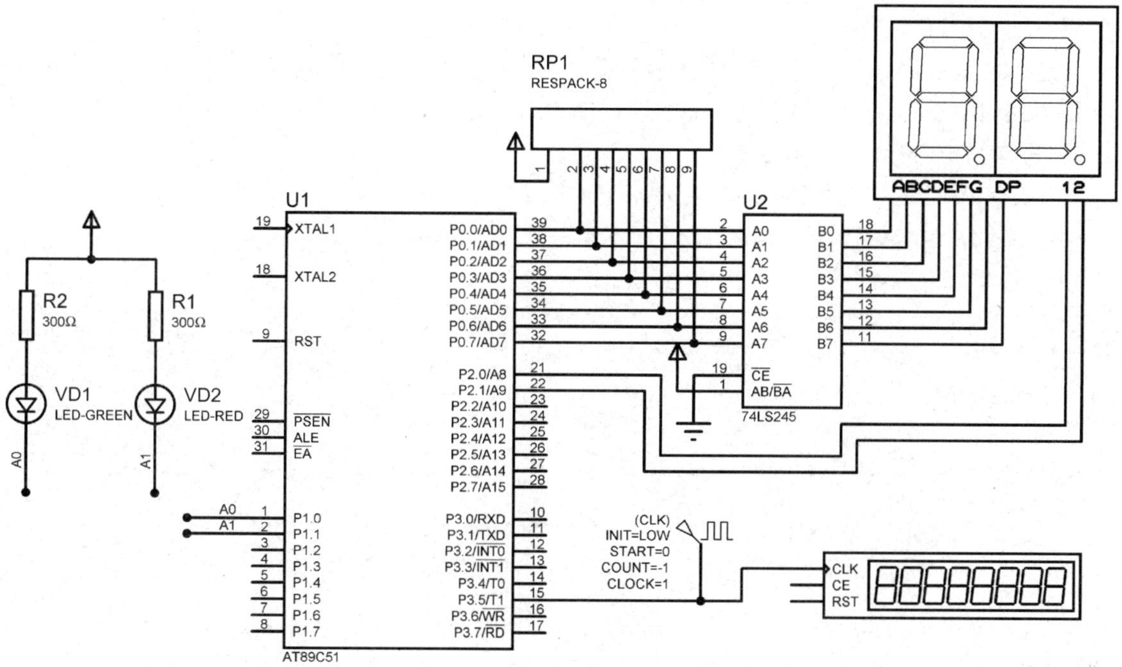

图 6-4 流水线计数系统电路图

【任务分析】

根据设计要求，采用两位一体共阴极数码管作为流水线计数系统的显示器件，采用计数器 T1 对外部脉冲（流水线物品）计数，虚拟仪表"COUNTER TIMER"实时显示外部脉冲的个数。

【实施步骤】

1. 添加元器件

打开 Proteus 仿真软件，按照表 6-1 添加元器件。

表 6-1 流水线计数系统的元器件清单

元器件名称	所属类	所属子类
AT89C51	Microprocessor ICs	8051 Family
RES	Resistors	Generic
7SEG-MPX2-CC	Optoelectronics	7-Segment Displays
74LS245	TTL 74LS series	Transceivers
RESPACK-8	Resistors	Resistor Packs
LED-RED	Optoelectronics	LEDs
LED-GREEN	Optoelectronics	LEDs

2. 绘制仿真图

元器件全部添加后，在 Proteus ISIS 的原理图编辑窗口中按图 6-4 绘制流水线计数系统仿真图。

3. 编写程序

设计要求 P3.5 口实现对外部脉冲计数，因此可以采用 T1 方式 2 来实现，TMOD=01100000B=0x60。由于数码管的计数值要求实时显示外部脉冲个数，可以将 T1 计数值设置为 1，所以 TH0=TL0=256-1=255。

在 Keil μVision5 中编写程序，实现流水线计数系统的效果，参考程序如下：

```
/************************************************************
程序名称：program6-2.c
程序功能：流水线计数系统程序
************************************************************/
#include<reg52.h>                         //加载头文件
#include<intrins.h>                        //加载头文件
/************************************************************
数据类型定义
************************************************************/
#define uchar unsigned char                //定义无符号字符型
#define uint unsigned int                   //定义无符号整型
/************************************************************
全局变量定义
************************************************************/
uchar COUNT=0;
/************************************************************
引脚定义
************************************************************/
sbit LED_G=P1^0;                          //红灯
sbit LED_R=P1^1;                          //绿灯
/************************************************************
数组定义：共阴极数码管的编码表
************************************************************/
uchar code LED_7SEG_CC[]=
{   //0      1     2     3     4     5     6     7     8     9
    0x3f, 0x06, 0x5b, 0x4f, 0x66, 0x6d, 0x7d, 0x07, 0x7f, 0x6f
};
/************************************************************
数组定义：显示单元，2 字节
************************************************************/
uchar DISPLAY_DATA[2]={0,0};
/************************************************************
函数名称：延时函数
功能描述：延时 t ms（晶振频率为 12MHz）
入口参数：t
************************************************************/
void delayms(uchar t)
{
    uchar i;
```

流水线计数系统
参考程序

```
            while(t--)
                for(i=96;i>0;i--);                      //实现 1ms 延时
}
/****************************************************************************
显示函数
****************************************************************************/
void display()
{
    uchar i,j=0xfe;
    for(i=0;i<2;i++)
    {
        P2=j;                                    //点亮一个数码管
        P0=LED_7SEG_CC[DISPLAY_DATA[i]];         //查表输出显示数据
        delayms(1);                              //延时 1ms
        j=_crol_(j,1);                           //准备点亮下一个数码管
    }
}
/****************************************************************************
函数名称：T1 中断函数
功能描述：T1 对外部脉冲计数函数
入口参数：无
****************************************************************************/
void ZD_T1() interrupt 3
{
    COUNT++;
    if(COUNT>=30)                                //判断计数值是否溢出
    {
        COUNT=30;                                //若溢出，则显示最大值
        LED_R=0;                                 //红灯亮
        LED_G=1;                                 //绿灯灭
    }
    DISPLAY_DATA[0]=COUNT/10;
    DISPLAY_DATA[1]=COUNT%10;
}
/****************************************************************************
主函数
****************************************************************************/
void main()
{
    LED_R=1;                                     //红灯灭
    LED_G=0;                                     //绿灯亮
    TMOD=0x60;                                   //T1 的方式 2
    TH1=255;                                     //T0 计满 1 个周期的初始值
    TL1=255;                                     //对外部脉冲计满 1 个周期产生中断
    TR1=1;                                       //开启 T1
```

```
        EA=1;                              //开启中断总开关
        ET1=1;                             //允许 T1 中断
        TR1=1;                             //T1 开启工作
    while(1)
        display();
}
```

4. 系统仿真

当 Keil C51 编译成功后，会自动产生 HEX 文件，接着打开之前绘制的 Proteus 仿真图，双击 AT89C51，弹出"Edit Component"对话框，单击"Program File"中的文件夹按钮，在弹出的"Select File Name"对话框中，选择之前编译生成的 HEX 文件，单击"打开"按钮，返回"Edit Component"对话框，单击"OK"按钮，即可装入 HEX 文件。

流水线计数系统
仿真效果视频

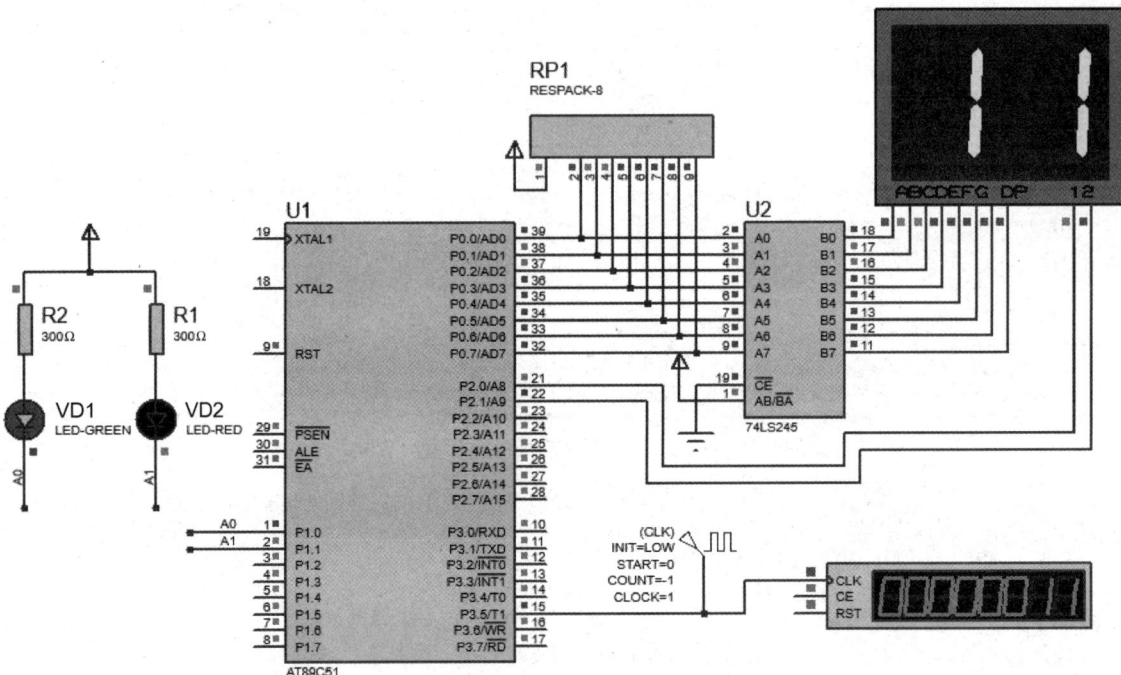

接着单击 Proteus ISIS 编辑界面左下角的运行按钮 ▶，即可观察是否能够实现流水线计数系统的显示效果，如图 6-5 所示。

图 6-5　流水线计数系统仿真效果图

任务 6.2　停车场计数系统的仿真设计

停车场计数系统
导学材料

✎ 学习目标

【知识目标】

（1）了解并掌握外部中断的原理。

（2）了解并掌握外部中断的使用方法。

【技能目标】

（1）了解并掌握单片机仿真软件 Proteus 的使用方法。

（2）了解并掌握单片机编译软件 Keil C51 的使用方法。

（3）通过停车场计数系统的仿真设计进一步掌握单片机项目的开发步骤。

【思政目标】

在停车场计数系统的仿真设计中，强调遵守交通规则、停车管理规定等的重要性，培养学生的规则意识和法治思维，并引导学生思考停车场计数系统如何与智慧城市、绿色出行等理念相结合，促进城市交通的可持续发展和环境保护。

6.2.1 外部中断的使用

1. TCON

在项目五中，介绍了 TCON 的高 4 位，而低 4 位为存放外部中断的触发方式控制位和锁存外部中断请求源，各位的定义如表 6-2 所示。

外部中断的使用
微课视频

表 6-2 TCON 低 4 位中各位的定义

位地址	8BH	8AH	89H	88H
位标志	IE1	IT1	IE0	IT0

（1）$\overline{INT0}$ 的中断触发方式选择位 IT0。

当 IT0=0 时，表示电平触发，低电平有效；当 IT0=1 时，表示边沿触发，下降沿有效。

（2）$\overline{INT0}$ 的中断请求标志位 IE0。

当 IT0=0，即低电平触发时，若 P3.2 口检测到低电平，则认为有中断请求，随即将 IE0 置位，向 CPU 申请中断。当 CPU 响应中断后，由硬件自动复位。

当 IT0=1，即下降沿触发时，若 P3.2 口检测到下降沿，则认为有中断请求，随即将 IE0 置位，向 CPU 申请中断。当 CPU 响应中断后，由硬件自动复位。

（3）$\overline{INT1}$ 的中断触发方式选择位 IT1。

其功能及操作情况同 IT0。

（4）$\overline{INT1}$ 的中断请求标志位 IE1。

其功能及操作情况同 IE0。

【课堂实训】编程实现 2 个按键控制数码管显示，如图 6-6 所示，每按动一次加按键，数码管显示值加 1，每按动一次减按键，数码管显示值减 1（加按键采用 $\overline{INT0}$ 低电平触发方式，减按键采用 $\overline{INT1}$ 下降沿触发方式）。

分析：题目要求加按键采用低电平触发方式，因此进入中断服务程序后需要判断按键是否被释放，而减按键采用下降沿触发方式，无须判断按键是否被释放。

课堂实训
仿真效果视频

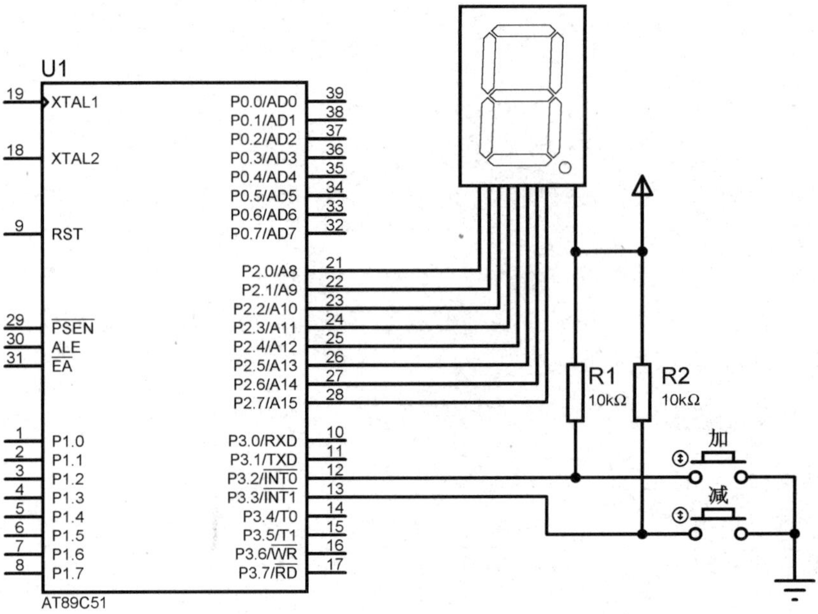

图 6-6　2 个按键控制数码管显示仿真效果图

参考程序如下：

```
/***********************************************************************

程序名称：program6-3.c
程序功能：按键控制数码管显示程序
***********************************************************************/
#include<reg52.h>                          //加载头文件
/***********************************************************************
引脚定义
***********************************************************************/
sbit KEY_ADD=P3^2;                         //加按键
/***********************************************************************
全局变量定义
***********************************************************************/
uchar COUNT=0;                             //计数值
/***********************************************************************
数组定义：共阳极数码管的编码表
***********************************************************************/
uchar LED7SEG_CA[]=
{
    //0     1     2     3     4     5     6     7     8     9
    0xc0, 0xf9, 0xa4, 0xb0, 0x99, 0x92, 0x82, 0xf8, 0x80, 0x90
};
/***********************************************************************
延时 t ms 函数
***********************************************************************/
void delayms(uchar t)
```

```
{
    uchar i;
    while(t--)
        for(i=96;i>0;i--);
}
/*****************************************************************************
INT0 中断服务程序，实现加按键功能
采用低电平触发方式
*****************************************************************************/
void ZD_IN0() interrupt 0
{
    delayms(20);                          //延时 20ms 去抖动
    if(KEY_ADD==0)                        //判断是否有按键被按下
    {
        while(KEY_ADD==0);               //判断 INT0 是否停止触发
        COUNT++;
        if(COUNT>=10)
            COUNT=0;
    }
}
/*****************************************************************************
INT1 中断服务程序，实现减按键功能
采用下降沿触发方式
*****************************************************************************/
void ZD_IN1() interrupt 2
{
    COUNT--;
    if(COUNT==255)
        COUNT=9;
}
/*****************************************************************************
主函数
*****************************************************************************/
void main()
{
    EA=1;                                //开启中断总开关
    EX0=1;                               //允许 INT0 中断
    EX1=1;                               //允许 INT1 中断
    IT0=0;                               //INT0 采用低电平触发方式
    IT1=1;                               //INT1 采用下降沿触发方式
    while(1)
        P2=LED7SEG_CA[COUNT];
}
```

2. 利用定时器扩展外部中断源

51 单片机有 2 个定时器/计数器，具有 2 个内部中断标志和外部计数引脚。将定时器设置为计数方式,计数初始值设定为满量程,一旦从外部计数引脚输入一个负跳变信号,计数器就加 1 产生溢出中断。把外部计数输入端 T0(P3.4)或 T1(P3.5)作为扩充中断源输入,该定时器的溢出中断标志及服务程序作为扩充中断源的标志及服务程序。

例如,将定时器 T0 设定为方式 2(自动重装载常数)代替一个扩充外部中断源,TH0 和 TL0 的初始值均为 0ffh,允许 T0 中断,CPU 开放中断,其初始化程序如下:

```
TMOD=0x06;
TH0=0xff;
TL0=0xff;
EA=1;
ET0=1;
TR0=1;
```

当连接在 T0(P3.4)引脚的外部中断请求输入线发生负跳变时,TL0 加 1 产生溢出,置位 TF0,向 CPU 发出中断请求。同时,TH0 的内容 0ffh 自动送到 TL0,即 TL0 恢复初始值。T0 引脚每输入 1 个负跳变信号,TF0 都会置位,且向 CPU 发出中断请求,相当于边沿触发的外部中断源输入。

6.2.2 停车场计数系统的任务实施

【设计要求】

如图 6-7 所示,停车场计数系统有 1 个入口和 1 个出口,在正常情况下,绿灯亮,红灯灭,表示停车场还有空余车位,当有车辆进来时,停车场内的车辆计数值加 1,当有车辆出去时,停车场内的车辆计数值减 1。当计数值超过 30 时,红灯亮,绿灯灭,表示停车场已停满,禁止车辆入内。

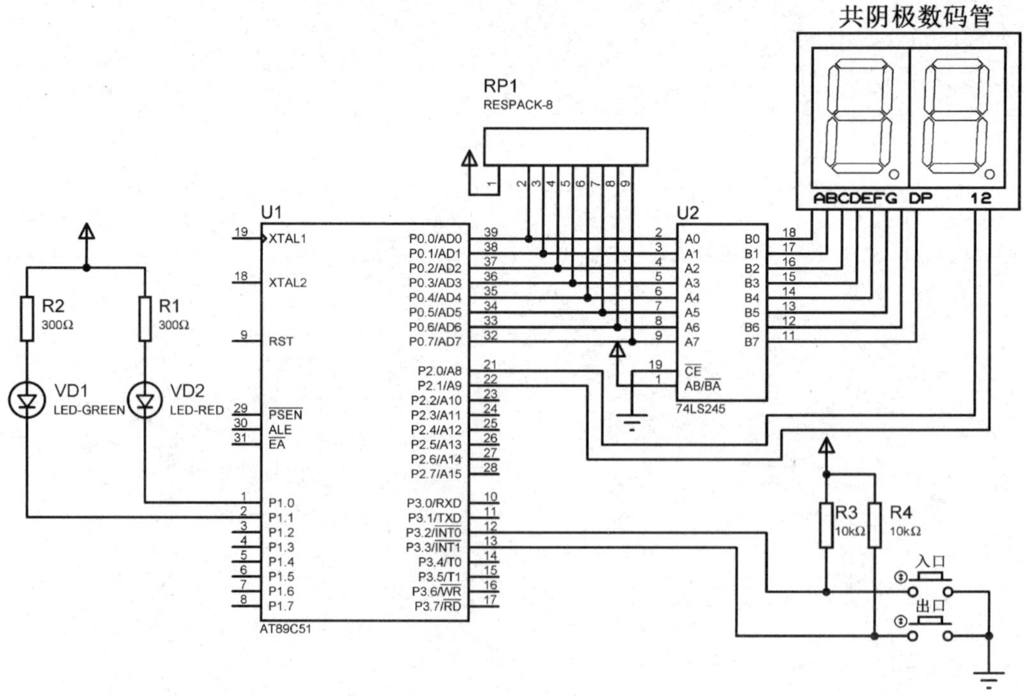

图 6-7 停车场计数系统电路图

【任务分析】

在本任务中，使用 2 个按键表示停车场入口和出口检测，每按下一次入口按键，表示一辆车驶入停车场，每按下一次出口按键，表示一辆车驶出停车场。入口按键接单片机的 P3.2 口，采用 $\overline{INT0}$ 进行检测；出口按键接单片机的 P3.3 口，采用 $\overline{INT1}$ 进行检测。

【实施步骤】

1. 添加元器件

打开 Proteus 仿真软件，按照表 6-3 添加元器件。注意：用 Proteus 仿真软件绘制单片机仿真图时，可以省略振荡电路和复位电路。

表 6-3　停车场计数系统的元器件清单

元器件名称	所属类	所属子类
AT89C51	Microprocessor ICs	8051 Family
RES	Resistors	Generic
74LS245	TTL 74LS series	Transceivers
RESPACK-8	Resistors	Resistor Packs
7SEG-MPX2-CC	Optoelectronics	7-Segment Displays
LED-RED	Optoelectronics	LEDs
LED-GREEN	Optoelectronics	LEDs
BUTTON	Switches & Relays	Switches

2. 绘制仿真图

元器件全部添加后，在 Proteus ISIS 的原理图编辑窗口中按图 6-7 绘制停车场计数系统仿真图。

3. 编写程序

在 Keil μVision5 中编写程序，实现停车场计数系统的效果，参考程序如下：

停车场计数系统
参考程序

```
/********************************************************************
程序名称：program6-4.c
程序功能：停车场计数系统程序
********************************************************************/
#include<reg52.h>                          //加载头文件
#include<intrins.h>                        //加载头文件
/********************************************************************
数据类型定义
********************************************************************/
#define uchar unsigned char                //定义无符号字符型
/********************************************************************
单片机引脚定义
********************************************************************/
sbit INPUT=P3^2;                           //入口引脚定义
sbit OUTPUT=P3^3;                          //出口引脚定义
sbit LED_GREEN=P1^1;                       //红灯
sbit LED_RED=P1^0;                         //绿灯
```

```
/********************************************************************
数组定义：共阴极数码管的编码表
********************************************************************/
uchar code LED_7SEG_CC[]=
{   //0     1     2     3     4     5     6     7     8     9
    0x3f, 0x06, 0x5b, 0x4f, 0x66, 0x6d, 0x7d, 0x07, 0x7f, 0x6f
};
/********************************************************************
全局变量定义
********************************************************************/
uchar COUNT=0;                                  //车辆计数值
uchar DISPLAY_DATA[]={0,0};                      //显示单元
/********************************************************************
延时 t ms 函数
********************************************************************/
void delayms(uchar t)
{
    uchar i;
    while(t--)
        for(i=96;i>0;i--);
}
/********************************************************************
显示函数
********************************************************************/
void display()
{
    uchar i,j=0xfe;
    for(i=0;i<2;i++)
    {
        P2=j;                                   //点亮一个数码管
        P0=LED_7SEG_CC[DISPLAY_DATA[i]];        //查表输出显示数据
        delayms(1);                             //延时 1ms
        j=_crol_(j,1);                          //准备点亮下一个数码管
    }
}
/********************************************************************
延时 20ms 去抖动函数
********************************************************************/
void delay()
{
    uchar i;
    for(i=10;i>0;i--)                           //10×2m=20ms
        display();
}
/********************************************************************
INT0 中断服务程序，实现对入口的检测
采用低电平触发方式
```

```
*********************************************************************/
void ZD_IN0() interrupt 0
{
    delay();                                //延时 20ms 去抖动
    if(INPUT==0)                            //判断是否有车辆驶入
    {
        while(INPUT==0)                     //判断 INT0 是否停止触发
            display();
        COUNT++;
        if(COUNT>=30)
        {
            COUNT=30;
            LED_RED=0;                      //红灯亮
            LED_GREEN=1;                    //绿灯灭
        }
    }
}
/*********************************************************************
INT1 中断服务程序，实现对出口的检测
采用下降沿触发方式
*********************************************************************/
void ZD_IN1() interrupt 2
{
    LED_RED=1;                              //红灯灭
    LED_GREEN=0;                            //绿灯亮
    COUNT--;
    if(COUNT==255)
        COUNT=0;
}
/*********************************************************************
主函数
*********************************************************************/
void main()
{
    LED_RED=1;                              //红灯灭
    LED_GREEN=0;                            //绿灯亮
    EA=1;                                   //开启中断总开关
    EX0=1;                                  //允许 INT0 中断
    EX1=1;                                  //允许 INT1 中断
    IT0=0;                                  // INT0 采用低电平触发方式
    IT1=1;                                  // INT1 采用下降沿触发方式
    while(1)
    {
        DISPLAY_DATA[0]=COUNT/10;
        DISPLAY_DATA[1]=COUNT%10;
        display();
    }
}
```

4．系统仿真

当 Keil C51 编译成功后，会自动产生 HEX 文件，接着打开之前绘制的 Proteus 仿真图，双击 AT89C51，弹出"Edit Component"对话框，单击"Program File"中的文件夹按钮，在弹出的"Select File Name"对话框中，选择之前编译生成的 HEX 文件，单击"打开"按钮，返回"Edit Component"对话框，单击"OK"按钮，即可装入 HEX 文件。

停车场计数系统
仿真效果视频

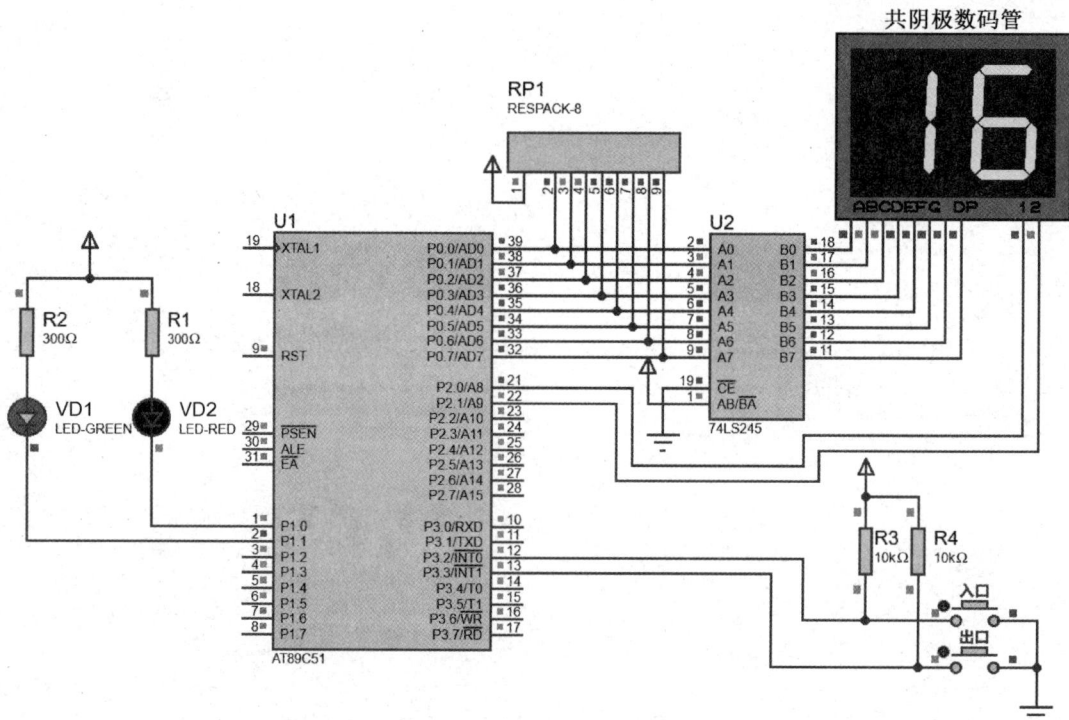

接着单击 Proteus ISIS 编辑界面左下角的运行按钮 ▶，即可观察是否能够实现停车场计数系统的功能，如图 6-8 所示。

图 6-8 停车场计数系统仿真效果图

素养小课堂

主流的单片机产品

8051 单片机最早由 Intel 公司推出，随后 Intel 公司将 80C51 单片机内核的使用权以专利互换的形式出让给芯片制造公司，如恩智浦、NEC、Atmel、AMD、Dallas、Siemens、Fujitsu、OKI、华邦、LG、STC 等。在保持与 8051 单片机兼容的基础上，这些公司融入了自身的优势，扩展了针对满足不同测控对象要求的外围电路，开发出了上百种功能各异的新产品。

1．AT89S 系列与 AVR 单片机

Atmel 公司生产的具有 Flash ROM 的增强型 51 系列单片机在市场上仍然十分流行，其中，AT89S 系列单片机十分活跃，它是 8 位 Flash 单片机，与 8051 单片机兼容，采用静态时钟模式。

AT90 系列单片机是 Atmel 公司在 20 世纪 90 年代推出的单片机，是增强精简指令集（RISC）结构、全静态工作方式、内载在线可编程 Flash 的单片机，也叫 AVR 单片机，与 PIC 单片机类似，其显著特点为高性能、高速度、低功耗。

AVR 单片机的型号较多，有 3 个档次：低档 Tiny 系列 AVR 单片机，主要有 Tiny11/12/13/15/25/28 等；中档 AT90S 系列 AVR 单片机，主要有 AT9051200/2313/8515/8535 等（正在淘汰或转型为 Mega）；高档 ATmega 系列 AVR 单片机，主要有 ATmega8/16/32/64/128（存储容量为 816/32/64/128，单位为 KB）和 ATmega8515/8535 等，开源电子原型平台 Arduino 采用的是 Mega 系列 AVR 单片机。

2. PIC 单片机

Microchip 的主要产品是 PIC 16F 系列、18F 系列的 8 位单片机，其突出特点是体积小、功耗低、精简指令集、运行速度快、抗干扰性好、可靠性高、有较强的模拟接口、代码保密性好、价格低、大部分芯片有兼容的 Flash ROM，适用于用量大、档次低、价格敏感的产品。

3. STC 单片机

STC 单片机是宏晶科技有限公司设计的基于 51 内核的国产单片机，目前在国内 8 位单片机市场上的占有率很高。STC 单片机指令集为复杂指令集，其优点是加密性强，很难解密或破解，具有超强的抗干扰性、功耗低、价格低，适用于各领域的设备控制。它的下载程序简单，在学校的教学中使用非常广泛，基本上取代了 Atmel 公司的 AT89/90 系列单片机。

4. 恩智浦单片机

恩智浦单片机有两个系列，一个是原飞利浦的 51LPC 系列，是基于 80C51 内核的单片机，嵌入了掉电检测、模拟及片内 RC 振荡器等功能，使之在高集成度、低成本、低功耗的应用设计中可以满足多方面的性能要求；另一个是飞思卡尔单片机，于 2015 年并入恩智浦。飞思卡尔单片机源于摩托罗拉半导体，主要应用在汽车、网络、工业、消费电子领域，在汽车电子领域占有较大的市场份额，单片机种类从 8 位覆盖到 32 位。

5. 德州仪器单片机

德州仪器提供了 TMS370 和 MSP430 两大系列的通用型单片机。TMS370 系列单片机是 8 位 CMOS 单片机，具有多种存储模式和外围接口模式，适用于复杂的实时控制场合；MSP430 系列单片机是一种功耗超低、功能集成度较高的 16 位单片机，特别适用于要求低功耗的场合。

6. STM 单片机

STM 单片机是意法半导体推出的系列单片机，拥有众多品种，从稳健的低功耗 8 位单片机 STM8 系列到基于 ARM Cortex-M0 和 M0+、Cortex-M3、Cortex-M4、Cortex-M7 内核的 32 位闪存微控制器 STM32 系列，为嵌入式产品开发人员提供了丰富的单片机资源。同时，意法半导体在不断扩大，并拓展其产品线，包括各种超低功耗单片机系列。

7. 英飞凌单片机

英飞凌的前身是 Siemens 的半导体部门。英飞凌 8 位单片机能实现高性能的电机驱动控制，在严酷环境（高温、EMI、振动）下具有极高的可靠性。英飞凌 8 位单片机主要有 XC800、XC886、XC888、XC82x、XC83x 等系列。英飞凌单片机多用于汽车、工业类产品，消费类产品应用较少。

8. 瑞萨单片机

瑞萨是由 NEC、三菱公司的半导体部门合并成立的，其单片机在汽车电子类市场占有较大的份额，而消费类占比很小。

9. 其他国产单片机

国产 8 位单片机的型号众多，涵盖多个品牌和系列，以满足不同领域和应用场景的需求。以下是一些具有代表性的国产 8 位单片机的型号及特点。

（1）中微半导体 SC8P 系列单片机。

中微半导体 SC8P 系列单片机是基于 Intel 8051 内核的 8 位单片机，具有高效的指令执行能力和很快的运算速度，主要代表型号包括 SC8P1151A、SC8P054AD、SC8P1712E、SC8P062AD 等。这些型号广泛应用于各类电子设备，如工业控制设备、家电产品等。

（2）华芯微 HC11 系列单片机。

华芯微 HC11 系列单片机兼容 Motorola MC68HC11 内核的 8 位单片机，在消费电子和工业控制等领域有着广泛的应用。华芯微还提供了 NuMicro 系列单片机，其中包含一些基于 8051 内核的 8 位产品，为用户提供了更多选择。

（3）北京君正集成电路 KX 系列单片机。

北京君正集成电路 KX 系列单片机是一款专为消费电子、工业控制等领域设计的 8 位单片机，该系列单片机以稳定的性能和适中的价格，在市场上获得了良好的口碑。

（4）复旦微电子 FD 系列单片机。

复旦微电子 FD 系列单片机是基于 CMOS 工艺的 8 位单片机，适用于多种嵌入式应用。该系列单片机以低功耗、高性能的特点受到用户的青睐。

（5）杭州士兰微电子 SM 系列单片机。

杭州士兰微电子 SM 系列单片机是一款性能与成本比极佳的 8 位单片机，在工业自动化、消费电子等领域有广泛的应用。其价格适中且性能稳定，是许多项目首选的 8 位单片机之一。

（6）新唐科技 N76 系列单片机。

新唐科技 N76 系列单片机是基于 8051 内核的 8 位单片机，具有高性能、低功耗和丰富的外设接口。该系列单片机适用于各种工业控制、家电、消费电子等领域。新唐科技还提供了完善的技术支持和强大的软件库，方便开发者快速实现项目开发。

（7）其他品牌和型号。

除了上述品牌，还有许多国产 8 位单片机型号可供选择。例如，炬微半导体的 8 位单片机在小家电、消费电子、安防等领域有着广泛的应用；上海兆芯公司的 ZC 系列单片机也以其卓越的性能和高度集成度受到好评。

课后任务

1. 在任务 6.1 的基础上添加功能，增加 2 个按键控制流水线计数系统的开启和停止，如图 6-9 所示。当系统开机后，停止指示灯亮，工作指示灯灭，数码管显示 0，计数器也显示 0；当按下开始按键后，停止指示灯灭，工作指示灯亮，计数器和数码管都开始显示计数值；若计数值达到最大值 M，则工作指示灯灭，停止指示灯亮，计数器和数码管的显示值都保持不变；若未达到最大值 M，则按下停止按键，计数器和数码管的显示值都保持不变，停止指示灯亮，工作指示灯灭。显示效果参见二维码。

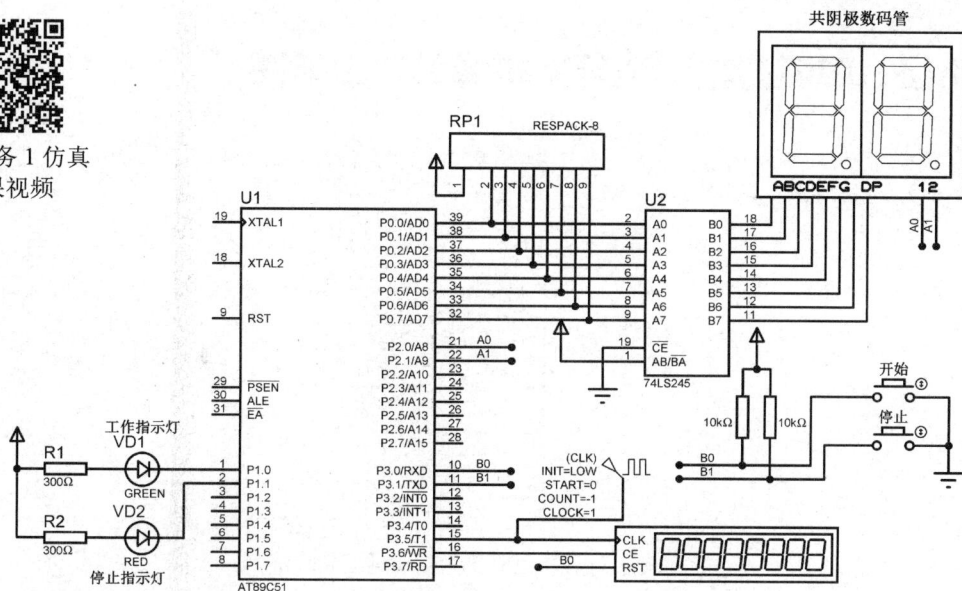

图 6-9　课后任务 1 电路图

2．在任务 6.2 的基础上添加功能，如图 6-10 所示。使得右边数码管显示停车场总车位数 50，左边数码管能够实时显示停车场已停车辆数，当已停车辆数等于总车位数时，红灯亮，绿灯灭，禁止车辆驶入停车场。显示效果参见二维码。

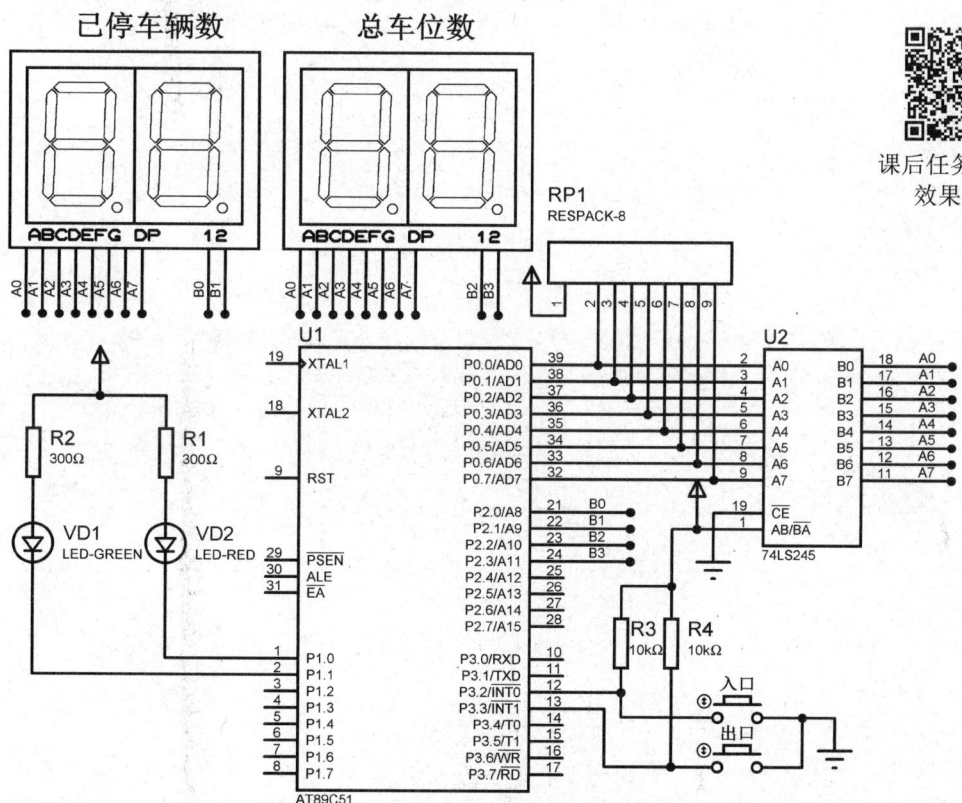

图 6-10　课后任务 2 电路图

知识拓展　中断和查询结合法扩展外部中断源

如果要扩展外部中断源，可以采用中断和查询结合的方法。图 6-11 所示为一种扩展外部中断源的实用方法。

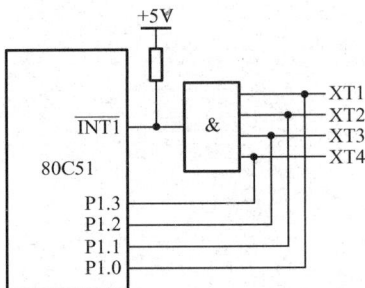

图 6-11　一种扩展外部中断源的实用方法

图 6-11 采用一个四与门扩展 4 个外部中断源 XT1～XT4，所有这些扩展的外部中断源都采用电平触发方式（低电平有效）。当 XT1～XT4 中有一个出现低电平时，与门输出为 0，使 $\overline{INT1}$ 为低电平触发中断。在 $\overline{INT1}$ 中断服务程序中，由软件按人为设定的顺序（优先级）查询外部中断源哪位为高电平，然后进入该中断进行处理。

在此方法中，各路输入的有效中断电平应该在 CPU 实际响应该中断源之前保持有效，并在该中断服务程序返回前取消。

$\overline{INT1}$ 中断服务程序如下：

```
void ZD_IN1() interrupt 2
{
    P1=0xff;
    switch(P1)
    {
        case 0xfe: XT1;break;        //当 XT1 按键被按下时，执行 XT1 函数
        case 0xfd: XT2;break;        //当 XT2 按键被按下时，执行 XT2 函数
        case 0xfb: XT3;break;        //当 XT3 按键被按下时，执行 XT3 函数
        case 0xf7: XT4;break;        //当 XT4 按键被按下时，执行 XT4 函数
    }
}
```

习题

一、单选题

1. $\overline{INT0}$ 允许中断的 C51 语句为（　　）。

A. RI=1;　　　　　　B. TR0=1;　　　　　　C. IT0=1;　　　　　　D. EX0=1;

2. 为使 P3.2 引脚出现的外部中断请求信号能得到 CPU 响应，必须满足的条件是（　　）。

A．ET0=1   B．EX0=1   C．EA=EX0=1   D．EA=ET0=1

3．80C51 单片机 $\overline{\text{INT1}}$ 和 $\overline{\text{INT0}}$ 的触发方式选择位是（   ）。

A．TR1 和 TR0   B．IE1 和 IE0   C．IT1 和 IT0   D．TF1 和 TF0

二、判断题

1．MCS-51 系列单片机在同级中断源同时申请中断时，CPU 首先响应 $\overline{\text{INT0}}$ 。（   ）

2．定时器 T0 中断可以被 $\overline{\text{INT0}}$ 中断。   （   ）

3．89C51 单片机的定时器/计数器是否工作可以通过外部中断进行控制。   （   ）

项目七　汉字点阵系统的仿真设计

任务 7.1　8×8 LED 点阵系统的仿真设计

8×8 LED 点阵
系统导学材料

学习目标

【知识目标】

（1）了解 8×8 LED 点阵显示器的结构和工作原理。

（2）了解并掌握 8×8 LED 点阵显示器的使用方法。

【技能目标】

（1）了解并掌握单片机仿真软件 Proteus 的使用方法。

（2）了解并掌握单片机编译软件 Keil C51 的使用方法。

（3）了解并掌握单片机程序下载的方法。

（4）了解并掌握单片机最小系统的组成。

（5）通过 8×8 LED 点阵系统的仿真设计初步了解并掌握单片机项目的开发步骤。

【思政目标】

在 8×8 LED 点阵系统的仿真设计中，鼓励学生发挥想象力，设计出新颖独特的显示效果，通过创新设计，激发学生的创造力和创新意识，培养学生的创新思维和实践能力。

7.1.1　8×8 点阵的使用

LED 点阵显示器一般应用于广告宣传、新闻传播等场合，其不仅能显示文字、图形、动画等，还可以分为单色显示和彩色显示。

8×8 点阵的使用
微课视频

1．LED 点阵显示器的结构

LED 点阵显示器是以发光二极管为像素组成阵列，用环氧树脂和塑膜封装而成的。采用这种封装的 LED 点阵显示器具有亮度高、引脚少、视角大、寿命长、耐湿、耐冷热且耐腐蚀等特点。

常见的 LED 点阵显示器按像素可分成 4×4、4×8、5×7、5×8、8×8、16×16、24×24、40×40 等不同规格。

一块 8×8 LED 点阵显示器实物图如图 7-1 所示，其对外共有 16 个引脚，引脚排列图如图 7-2 所示。8×8 LED 点阵显示器内部由 8 行 8 列的 LED 构成，其中，8 根行线用 H1～H8（行 1～行 8）表示，8 根列线用 L1～L8（列 1～列 8）表示，如图 7-3 所示。

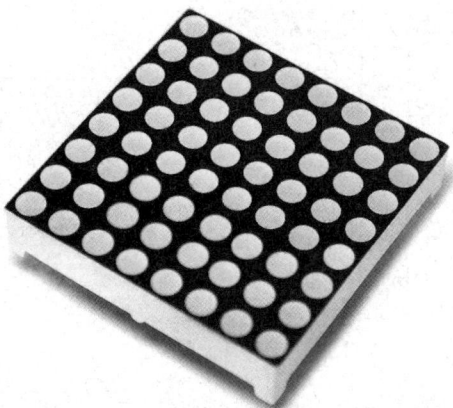

图 7-1 一块 8×8 LED 点阵显示器实物图

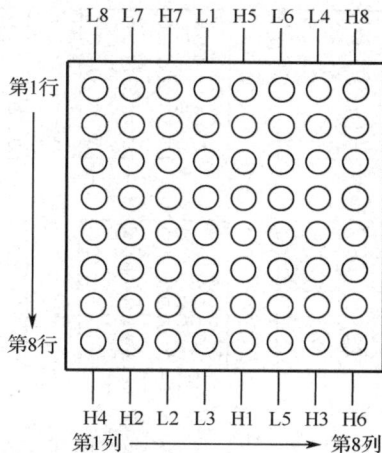

图 7-2 8×8 LED 点阵显示器引脚排列图

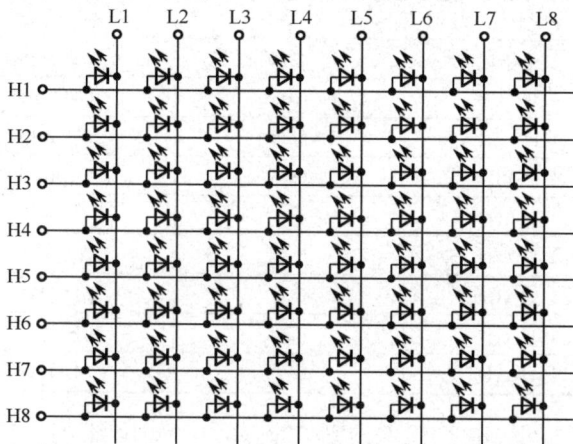

图 7-3 8×8 LED 点阵显示器内部结构图

从图 7-3 中可以看出，点亮跨接在某行某列上的 LED 的条件是：对应行为高电平，对应列为低电平。例如，当 H1=1，L1=0 时，左上角的 LED 点亮。

2. LED 点阵显示器的显示方式

LED 点阵显示器有两种显示方式：静态显示和动态显示。

静态显示原理简单、控制方便，但硬件接线复杂、占用 I/O 接口线较多。

在实际应用中一般采用动态显示，动态显示采用扫描的方式工作，从上到下逐行（或逐列）选通（点亮），同时向各列（或各行）送出表示显示信息的数据代码，循环往复，利用人眼的视觉暂留现象，将连续的几帧画面高速循环显示，只要帧率高于 24 帧/秒，人眼看到的就是一个完整的、连续的画面。

用动态扫描的方式显示汉字"丽"的过程如图 7-4 所示，采用列扫描方式，当某列输入低电平（工作）时，其余列输入高电平（不工作），然后输入行数据，若某行为高电平，则对应行线和列线交叉点位置的 LED 将被点亮，若某行为低电平，则对应的 LED 熄灭。汉字"丽"的行列编码如表 7-1 所示。

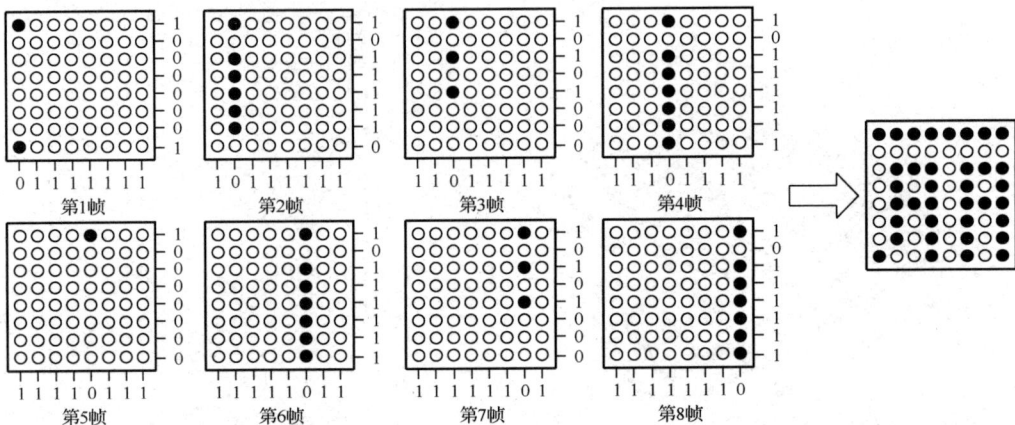

图 7-4 用动态扫描的方式显示汉字"丽"的过程

表 7-1 汉字"丽"的行列编码（列扫描方式）

扫描方式	列数据		行数据	
（列扫描）	二进制数	十六进制数	二进制数	十六进制数
第 1 帧	1111 1110	0xfe	1000 0001	0x81
第 2 帧	1111 1101	0xfd	1011 1110	0xbe
第 3 帧	1111 1011	0xfb	1010 1000	0xa8
第 4 帧	1111 0111	0xf7	1011 1111	0xbf
第 5 帧	1110 1111	0xef	1000 0000	0x80
第 6 帧	1101 1111	0xdf	1011 1111	0xbf
第 7 帧	1011 1111	0xbf	1010 1000	0xa8
第 8 帧	0111 1111	0x7f	1011 1111	0xbf

【课堂实训】编程实现 8×8 LED 点阵的显示，采用红色 LED 点阵（仿真模型为 MATRIX-8X8-RED）列扫描方式进行显示，电路图如图 7-5 所示，要求 8×8 LED 点阵显示汉字"丽"。

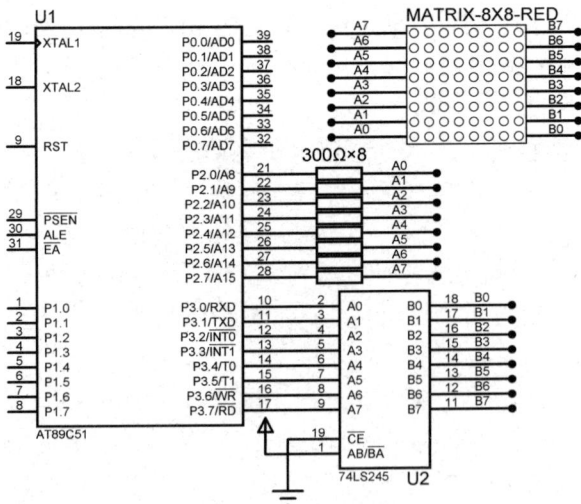

课堂实训
仿真效果视频

图 7-5 8×8 LED 点阵电路图

分析：题目要求采用红色 LED 点阵列扫描方式，其中 P3 口控制列线，P2 口控制行线，在每条行线上串接一个 300Ω 的限流电阻。同时，为提高单片机接口带负载的能力，P3 口通过 74LS245 与 LED 点阵连接，增大了 P3 口输出的电流，既保证了 LED 点阵的亮度，又保护了单片机接口引脚。

注意：在绘制仿真图时，需要将红色 LED 点阵逆时针旋转 90°。

参考程序如下：

```c
/*********************************************************************
程序名称：program7-1.c
程序功能：8×8 LED 点阵显示汉字"丽"程序
*********************************************************************/
#include<reg52.h>                              //加载头文件
#include<intrins.h>                            //加载头文件
/*********************************************************************
数据类型定义
*********************************************************************/
#define uchar unsigned char                    //定义无符号字符型
/*********************************************************************
数组定义：8×8 LED 点阵显示汉字"丽"的编码表
*********************************************************************/
uchar code LED_8X8[]=
{
    0x81, 0xbe, 0xa8, 0xbf, 0x80, 0xbf, 0xa8, 0xbf    //汉字"丽"的编码
};
/*********************************************************************
延时 t ms 函数
*********************************************************************/
void delayms(uchar t)
{
    uchar i;
    while(t--)
        for(i=96;i>0;i--);
}
/*********************************************************************
显示函数
*********************************************************************/
void display()
{
    uchar i,j=0xfe;
    for(i=0;i<8;i++)                            //共扫描 8 列
    {
        P3=j;                                  //点亮第 1 列
        P2=LED_8X8[i];                         //查表输出列数据
        j=_crol_(j,1);                         //左移
```

```
        delayms(1);                                      //延时 1ms
    }
}
/**************************************************************************
主函数
**************************************************************************/
void main()
{
    while(1)
        display();                                       //显示函数
}
```

7.1.2　8×8 LED 点阵系统的任务实施

【设计要求】

使用单片机和 2 块 8×8 LED 点阵设计一个汉字点阵显示系统，要求 8×8 LED 点阵显示"中华"两个汉字，电路图如图 7-6 所示，采用绿色的 8×8 LED 点阵。

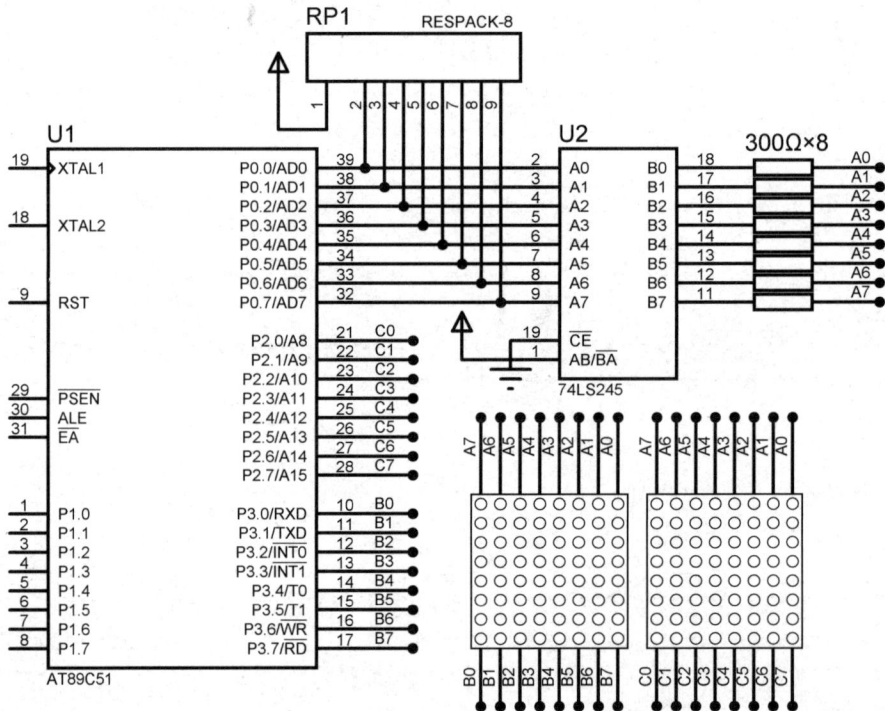

图 7-6　8×8 LED 点阵系统电路图

【任务分析】

根据设计要求，采用 2 块绿色的 8×8 LED 点阵作为显示器件，采用列扫描方式。由于在 Proteus 仿真软件中，绿色点阵与红色点阵的内部结构不同，因此需要采用列扫描共阳极的驱动方式进行扫描，当行数据为高电平时，LED 不亮；为低电平时，LED 亮。

【实施步骤】

1. 添加元器件

打开 Proteus 仿真软件，按照表 7-2 添加元器件。

表 7-2 8×8 LED 点阵系统的元器件清单

元器件名称	所属类	所属子类
AT89C51	Microprocessor ICs	8051 Family
RES	Resistors	Generic
MATRIX-8X8-GREEN	Optoelectronics	Dot Matrix Displays
74LS245	TTL 74LS series	Transceivers
RESPACK-8	Resistors	Resistor Packs

2. 绘制仿真图

元器件全部添加后，在 Proteus ISIS 的原理图编辑窗口中按图 7-6 绘制 8×8 LED 点阵系统仿真图。

3. 编写程序

由于系统采用 2 块 8×8 LED 点阵，因此显示函数包含两部分，一部分是 P3 口和 P0 口驱动显示汉字"中"，另一部分是 P2 口和 P0 口驱动显示汉字"华"。

在 Keil μVision5 中编写程序，实现 8×8 LED 点阵系统的效果，参考程序如下：

```
/**********************************************************
程序名称：program7-2.c
程序功能：8×8 LED 点阵系统程序
**********************************************************/
#include<reg52.h>                          //加载头文件
#include<intrins.h>                        //加载头文件
/**********************************************************
数据类型定义
**********************************************************/
#define uchar unsigned char                //定义无符号字符型
/**********************************************************
数组定义：8×8 LED 点阵显示汉字"中华"的编码表
**********************************************************/
uchar code LED_8X8[]=
{
    0xc7, 0xd7, 0xd7, 0x00, 0xd7, 0xd7, 0xc7, 0xff,    //汉字"中"的编码
    0xdb, 0x8b, 0x7b, 0xf0, 0x0b, 0xcb, 0xab, 0xff    //汉字"华"的编码
};
/**********************************************************
延时 t ms 函数
**********************************************************/
void delayms(uchar t)
{
```

8×8 LED 点阵系统
参考程序

```
        uchar i;
        while(t--)
                for(i=96;i>0;i--);
}
/************************************************************
显示函数
************************************************************/
void display()
{
        uchar i,j=0x01;
        P2=0x00;                                //P2 口不扫描
        for(i=0;i<8;i++)                        //显示汉字"中"
        {
                P3=j;                           //扫描 P3 口，点亮第 1 列
                P0=LED_8X8[i];                  //查表输出列数据
                j=_crol_(j,1);                  //左移
                delayms(1);                     //延时 1ms
        }
        P3=0x00;                                //P3 口不扫描
        for(i=0;i<8;i++)                        //显示汉字"华"
        {
                P2=j;                           //扫描 P2 口，点亮第 1 列
                P0=LED_8X8[i+8];                //查表输出列数据
                j=_crol_(j,1);                  //左移
                delayms(1);                     //延时 1ms
        }
}
/************************************************************
主函数
************************************************************/
void main()
{
        while(1)
                display();                      //显示函数
}
```

4. 系统仿真

当 Keil C51 编译成功后，会自动产生 HEX 文件，接着打开之前绘制的 Proteus 仿真图，双击 AT89C51，弹出"Edit Component"对话框，单击"Program File"中的文件夹按钮，在弹出的"Select File Name"对话框中，选择之前编译生成的 HEX 文件，单击"打开"按钮，返回"Edit Component"对话框，单击"OK"按钮，即可装入 HEX 文件。

8×8 LED 点阵系统
仿真效果视频

注意：为了更好地显示仿真效果，可以选择 Proteus 仿真软件的"System"菜单栏中的"Set Animation Options"子菜单，弹出"Animated Circuits Configuration"对话框，取消勾选"Animation

Options"选区中的 2 个复选框，如图 7-7 所示。

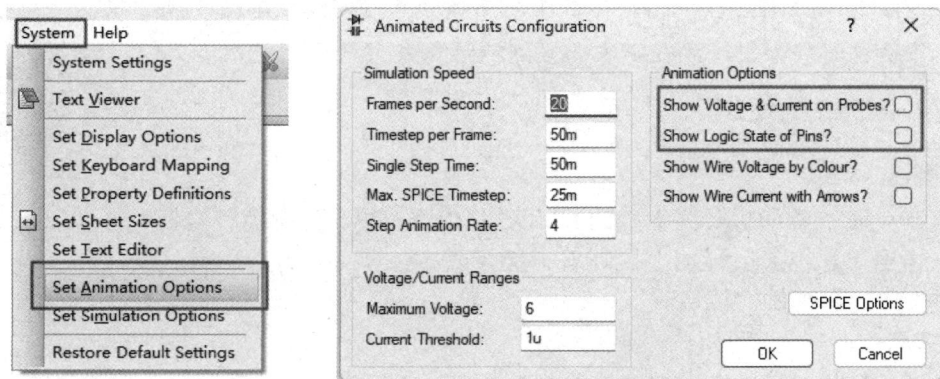

图 7-7 "Animated Circuits Configuration"对话框

接着单击 Proteus ISIS 编辑界面左下角的运行按钮▶，即可观察是否能够实现 8×8 LED 点阵系统的显示效果，如图 7-8 所示。

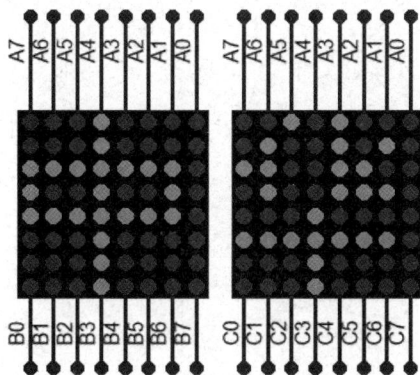

图 7-8 8×8 LED 点阵系统仿真效果图

任务 7.2 16×16 汉字点阵系统的仿真设计

学习目标

16×16 汉字点阵系统
导学材料

【知识目标】

（1）了解并掌握 74HC595 的引脚及功能。

（2）了解并掌握 16×16 汉字点阵的驱动电路和程序设计方法。

【技能目标】

（1）了解并掌握单片机仿真软件 Proteus 的使用方法。

（2）了解并掌握单片机编译软件 Keil C51 的使用方法。

（3）通过 16×16 汉字点阵系统的仿真设计进一步掌握单片机项目的开发步骤。

【思政目标】

在 16×16 汉字点阵系统的仿真设计中，培养学生认识汉字中蕴含的美感，结合北京奥运会的"击缶倒计时"节目，激发学生的爱国情结和人文素养。

7.2.1 74HC595 的使用

74HC595 芯片的使用微课视频

1. 概述

74HC595 是一款漏极开路输出的 CMOS 移位寄存器，输出接口为可控的三态输出端，也能串行输出控制下一级级联芯片，可作为数码管、点阵的显示驱动芯片。

2. 特点

（1）高速移位时钟频率，允许最大时钟频率为 25MHz。

（2）标准串行（SPI）接口。

（3）CMOS 串行输出，可用于多个设备的级联。

（4）低功耗，在 25℃下，电流 I_{CC} 最大值为 4μA。

3. 引脚排列图及功能说明

74HC595 共有 16 个引脚，其引脚排列图如图 7-9 所示，引脚功能说明如表 7-3 所示。

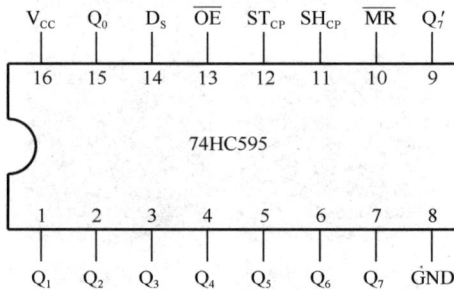

图 7-9　74HC595 的引脚排列图

表 7-3　74HC595 的引脚功能说明

引脚号	引脚名称	功能说明	引脚号	引脚名称	功能说明
1	Q_1	三态输出端	9	Q_7'	串行数据输出端
2	Q_2	三态输出端	10	$\overline{MR}$	移位寄存器清零端
3	Q_3	三态输出端	11	SH_{CP}	数据输入时钟端
4	Q_4	三态输出端	12	ST_{CP}	输出锁存器锁存时钟端
5	Q_5	三态输出端	13	$\overline{OE}$	输出使能端
6	Q_6	三态输出端	14	D_S	数据输入端
7	Q_7	三态输出端	15	Q_0	三态输出端
8	GND	地	16	V_{CC}	电源端

4. 真值表及时序图

74HC595 的真值表如表 7-4 所示，时序图如图 7-10 所示。

表 7-4　74HC595 的真值表

输入					输出
D_S	SH_{CP}	$\overline{MR}$	ST_{CP}	$\overline{OE}$	
×	×	×	×	×	$Q_0\sim Q_7$ 输出高阻
×	×	×	×	L	$Q_0\sim Q_7$ 输出有效值
×	×	L	×	×	移位寄存器清零
L	↑	H	×	×	移位寄存器存储低电平
H	↑	H	×	×	移位寄存器存储高电平
×	↓	H	×	×	移位寄存器状态保持
×	×	×	↑	×	输出存储器锁存移位寄存器中的状态值
×	×	×	↓	×	输出存储器状态保持

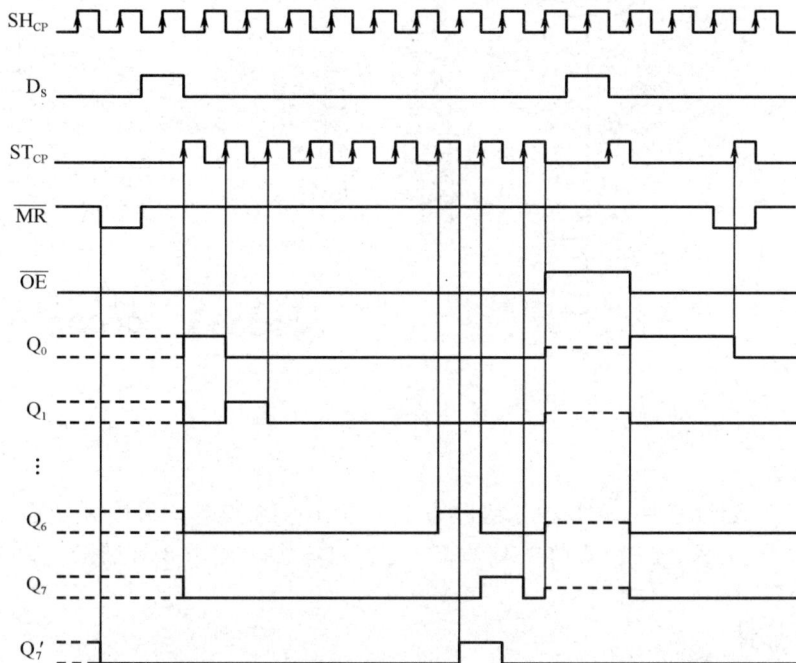

图 7-10　74HC595 的时序图

7.2.2　16×16 汉字点阵系统的任务实施

【设计要求】

如图 7-11 所示，通过单片机和 5 块 74HC595 设计一个 16×16 汉字点阵系统，并能够显示汉字"欢"，其中 16×16 点阵由 4 块 8×8 LED 点阵拼接而成。

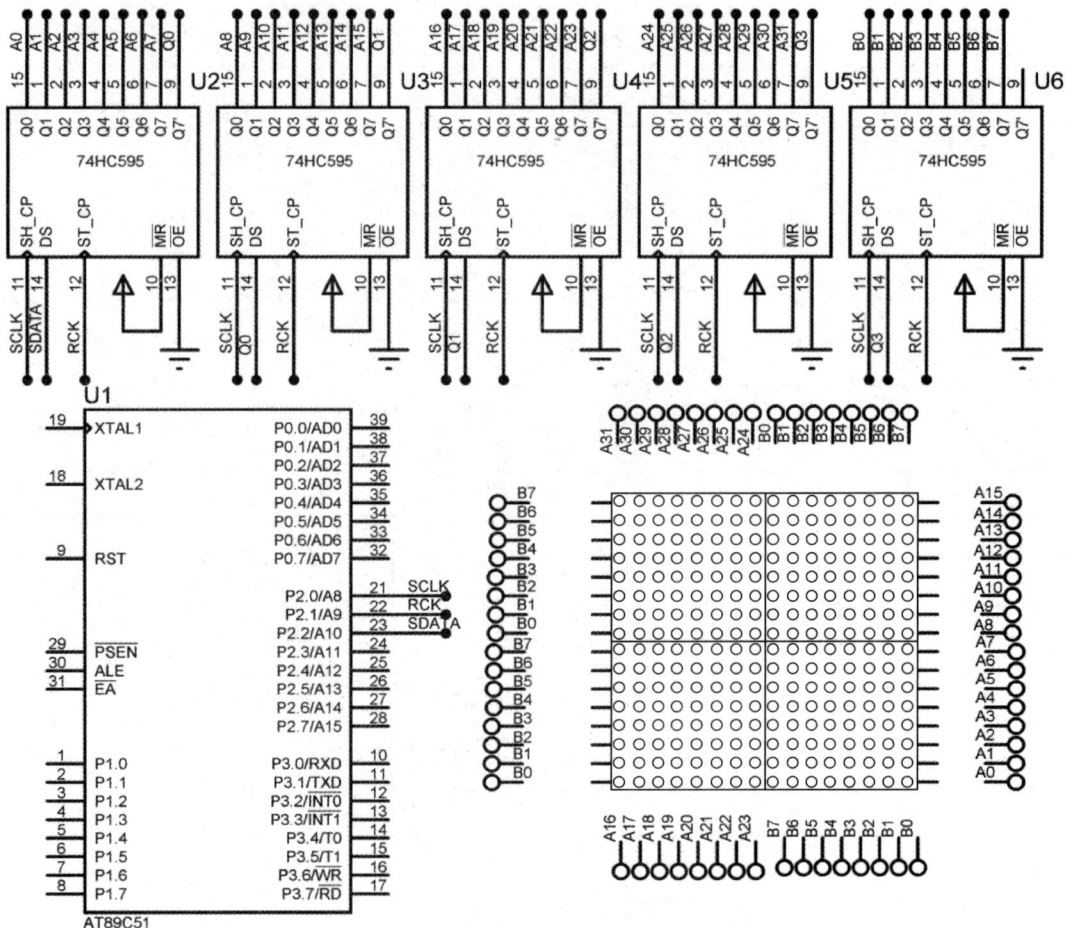

图 7-11　16×16 汉字点阵系统电路图

【任务分析】

在本任务中，通过 5 块 74HC595 级联方式驱动 16×16 点阵，采用列扫描方式，其中 U6 输出的数据送给 4 块 8×8 LED 点阵的列线端，U2～U5 输出 4 个 8×8 LED 点阵的行数据。

【实施步骤】

1. 添加元器件

打开 Proteus 仿真软件，按照表 7-5 添加元器件。注意：用 Proteus 仿真软件绘制单片机仿真图时，可以省略振荡电路和复位电路。

表 7-5　16×16 汉字点阵系统的元器件清单

元器件名称	所属类	所属子类
AT89C51	Microprocessor ICs	8051 Family
MATRIX-8X8-GREEN	Optoelectronics	Dot Matrix Displays
74HC595	TTL 74HC series	Registers

2. 绘制仿真图

元器件全部添加后，在 Proteus ISIS 的原理图编辑窗口中按图 7-11 绘制 16×16 汉字点阵系统仿真图。在绘制 16×16 点阵显示器时，首先将每块绿色的 8×8 LED 点阵逆时针旋转 90°，然后标记好每个引脚的网络标号，如图 7-12 所示。

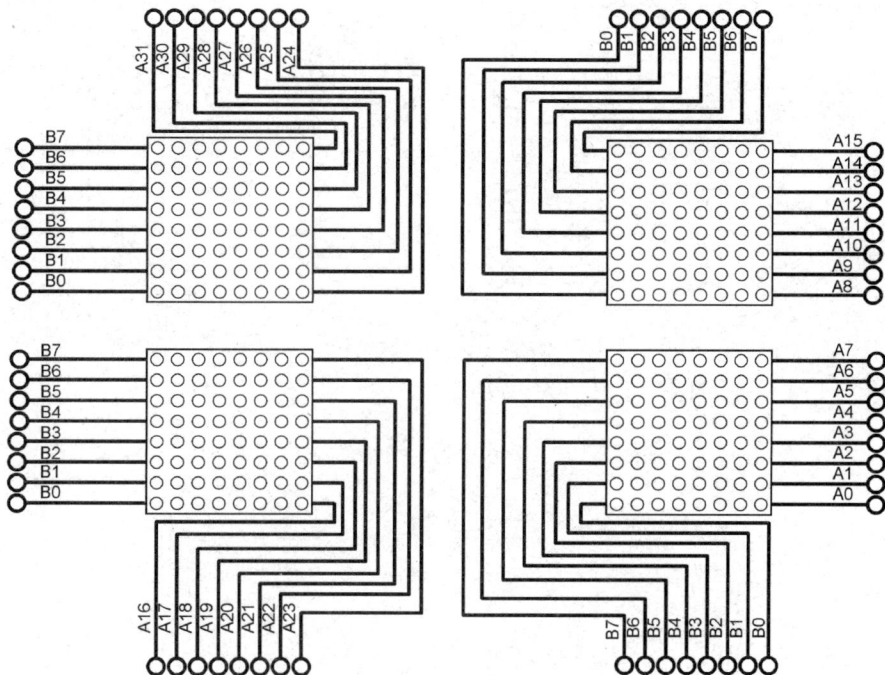

图 7-12　16×16 点阵分块绘制

接着双击 LED 点阵引脚的导线，将导线类型设置为"none"，如图 7-13 所示，这样就可以隐藏导线。

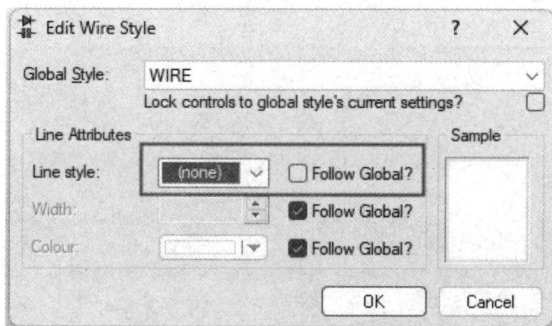

图 7-13　导线类型设置

最后将 4 块 8×8 LED 点阵拼接在一起，如图 7-14 所示。

3. 编写程序

在 Keil μVision5 中编写程序，实现 16×16 汉字点阵系统的效果。

目前已经有许多字模软件可以生成 16×16 汉字点阵的汉字编码。图 7-15 所示为一款 LED 点阵字模软件，输入所需显示的汉字，选择数据转换格式及语言输出格式，单击"转换"按钮，就可以直接生成所需显示汉字的编码。

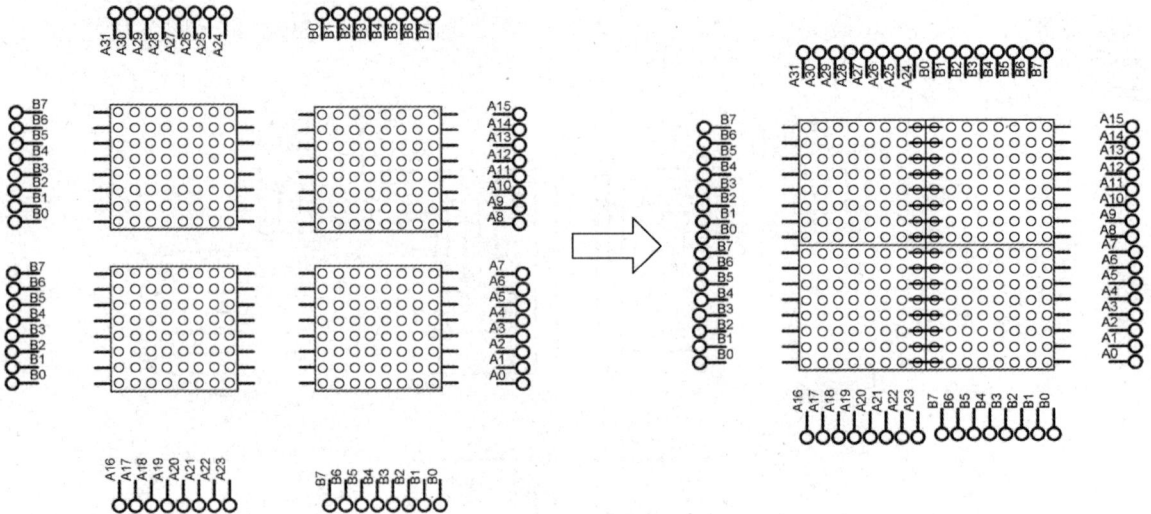

图 7-14　16×16 汉字点阵系统成型图

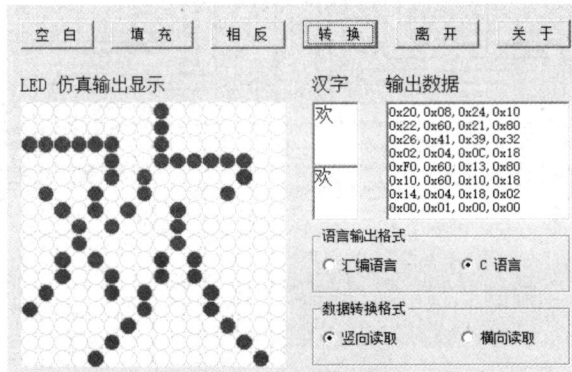

图 7-15　一款 LED 点阵字模软件

参考程序如下：

```
/*****************************************************
程序名称：program7-3.c
程序功能：16×16 汉字点阵系统程序
****************************************************/
#include<reg52.h>                          //加载头文件
#include<intrins.h>                         //加载头文件
/*****************************************************
数据类型定义
****************************************************/
#define uchar unsigned char                 //定义无符号字符型
/*****************************************************
单片机引脚定义
****************************************************/
sbit SCLK_595=P2^0;                         //移位时钟脉冲
```

16×16 汉字点阵系统
参考程序

```
sbit RCK_595=P2^1;                          //输出锁存器控制脉冲
sbit SDATA_595=P2^2;                        //串行数据输入
/*************************************************************
数组定义：16×16 汉字点阵系统的编码表，显示汉字"欢"
*************************************************************/
uchar code LED8X8_CODE[]=
{
0x20,  0x08,  0x24,  0x10,  0x22,  0x60,  0x21,  0x80,
0x26,  0x41,  0x39,  0x32,  0x02,  0x04,  0x0C,  0x18,
0xF0,  0x60,  0x13,  0x80,  0x10,  0x60,  0x10,  0x18,
0x14,  0x04,  0x18,  0x02,  0x00,  0x01,  0x00,  0x00
};
/*************************************************************
延时 1ms 函数，定时器 T0 产生
*************************************************************/
void delay1ms()
{
    TH0=0xfc;                               //12MHz，1ms 定时器初始值
    TL0=0x18;
    TR0=1;                                  //开启定时器
    while(TF0==0);                          //查询判断有无溢出
    TF0=0;                                  //若溢出，则清零
    TR0=0;                                  //关闭定时器
}
/*************************************************************
74HC595 写数据函数
*************************************************************/
void write_74hc595(uchar i)
{
    uchar j;
    for(j=8;j>0;j--)
    {
        i=_crol_(i,1);                      //左移位
        SDATA_595=(bit)(i&0x01);            //移位输出数据
        SCLK_595=1;                         //上升沿发生移位
        _nop_();
        _nop_();
        SCLK_595=0;
    }
}
/*************************************************************
74HC595 驱动函数
*************************************************************/
void ini_74hc595(uchar i,j)
{
```

```
        write_74hc595(j);                          //输出列数据
        write_74hc595(LED8X8_CODE[i]);             //查表输出左上方点阵数据
        write_74hc595(LED8X8_CODE[i+1]);           //查表输出左下方点阵数据
        write_74hc595(LED8X8_CODE[i+16]);          //查表输出右上方点阵数据
        write_74hc595(LED8X8_CODE[i+17]);          //查表输出右下方点阵数据
        RCK_595=0;
        _nop_();
        _nop_();
        RCK_595=1;                                 //上升沿将数据送到输出锁存器
        _nop_();
        _nop_();
        _nop_();
        RCK_595=0;
}
/********************************************************************
显示函数，占用 8ms
********************************************************************/
void display()
{
    uchar i,j=0xfe,k=0;
    for(i=0;i<8;i++)                               //扫描 8 次
    {
        ini_74hc595(k,j);                          //74HC595 驱动函数
        delay1ms();                                //延时 1ms
        j=_crol_(j,1);
        k=k+2;
    }
}
/********************************************************************
主函数
********************************************************************/
void main()
{
    TMOD=0x01;                                     //定时器 T0 的方式 1
    while(1)
        display();                                 //调用显示函数，占用 8ms
}
```

4. 系统仿真

当 Keil C51 编译成功后，会自动产生 HEX 文件，接着打开之前绘制的 Proteus 仿真图，双击 AT89C51，弹出"Edit Component"对话框，单击"Program File"中的文件夹按钮，在弹出的"Select File Name"对话框中，选择之前编译生成的 HEX 文件，单击"打开"按钮，返回"Edit Component"对话框，单击"OK"按钮，即可装入 HEX 文件。

16×16 LED 汉字点阵
系统仿真效果视频

接着单击 Proteus ISIS 编辑界面左下角的运行按钮 ▶，即可观察是否能够实现 16×16 汉字点阵系统的功能，如图 7-16 所示。

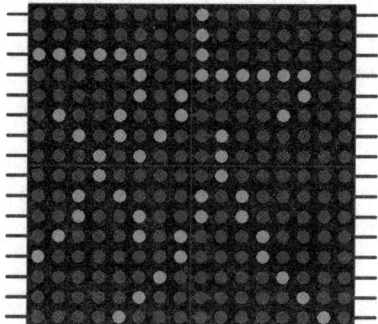

图 7-16　16×16 汉字点阵系统仿真效果图

素养小课堂

2008 年北京奥运会开幕式的"击缶倒计时"节目

2008 年 8 月 8 日，那是一个被历史铭记的夜晚，北京奥运会开幕式在国家体育场——"鸟巢"隆重举行。这一夜，不仅是中国人的骄傲，还是全球体育迷的盛宴。而开幕式中的"击缶倒计时"节目，更是以其独特的创意、精湛的表演和深厚的文化底蕴，成为整场开幕式的点睛之笔，给全世界留下了深刻而难忘的印象。

1. 震撼人心的开场

随着夜幕降临，"鸟巢"内灯火辉煌，9 万多观众席无虚席，每个人的脸上都洋溢着期待与激动的笑容。突然，一道耀眼的天光照亮了古老的日晷，也照亮了全场的目光。随着日晷强光的反射，2008 个身着银衣的演员手持特制的缶，排列成整齐的方阵，出现在观众眼前。他们的动作整齐划一，仿佛经过长时间的精心排练，每个细节都透露出严谨与专注。

2. 创意无限的击缶倒计时

随着一声令下，演员们开始有节奏地击打手中的缶。每次击打都伴随着响亮的声音和璀璨的光芒，仿佛是在向世界宣告这一历史时刻的到来。缶面上白色的灯光依次闪亮，组合出倒计时的数字，从 60 到 1，每个数字都牵动着全球观众的心弦。

这不仅仅是一个简单的倒计时过程，更是一次中华文化的展示。缶作为中国古代的传统打击乐器，其悠久的历史和深厚的文化底蕴在这一刻得到了充分体现。而"击缶倒计时"节目巧妙地将其与现代科技相结合，每个缶就代表一个发光二极管，缶被击打表示发光二极管点亮，缶未被击打，表示发光二极管熄灭，通过不同的组合实现不同的显示效果，从而构成"击缶点阵屏"，使得这一传统乐器——缶，焕发出了新的生命力。

3. 精准无误的表演

在"击缶倒计时"节目中，演员们的表演堪称完美无瑕。他们的动作整齐划一，每次击打都准确无误地落在了缶面上。这种精准度不仅来自长时间的排练和严格的训练，更来自演员们对中华文化的热爱和尊重。

在表演过程中，演员们的脸上始终保持着微笑和自信。他们的眼神中充满了对未来的期待和对祖国的自豪。这种精神风貌不仅感染了现场的观众，还通过电视屏幕传遍了全球。"击缶倒计时"节目虽然只有短短的几十秒时间，但它却给全球留下了深刻而持久的印象。它让全球看到了中华文化的魅力和活力，看到了中国人民的团结和自信，也向全球展示了中国文化的博大精深和与时俱进的创新精神。这种创新不仅体现在技术上，还体现在对传统文化的传承和发扬上。

课后任务

1. 在任务 7.1 的基础上添加功能，其电路图如图 7-6 所示。要求第 1s 显示"中华"两个汉字，第 2 秒显示"人民"两个汉字，重复循环。显示效果参见二维码。

2. 在任务 7.2 的基础上添加功能，其电路图如图 7-11 所示。要求 16×16 汉字点阵轮流显示"我爱中华"四个汉字，每个汉字显示 1s。显示效果参见二维码。

课后任务 1 仿真效果视频   课后任务 2 仿真效果视频

知识拓展　5×7 LED 点阵显示器

5×7 LED 点阵系统
仿真效果视频

一块 5×7 LED 点阵显示器实物图如图 7-17 所示，其内部由 7 行 5 列的 LED 构成。其中，7 根行线用 H1～H7（行 1～行 7）表示，5 根列线用 L1～L5（列 1～列 5）表示，内部结构如图 7-18 所示。

图 7-17　一块 5×7 LED 点阵显示器实物图

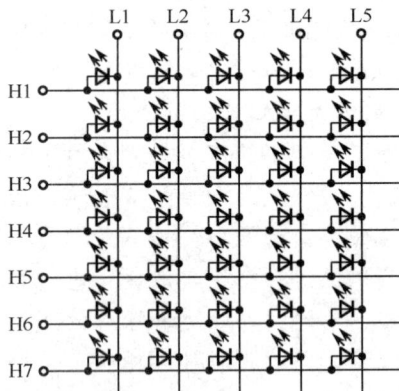

图 7-18　5×7 LED 点阵显示器的内部结构

从图 7-18 中可以看出，点亮跨接在某行某列的 LED 的条件是：对应行为高电平，对应列为低电平。例如，当 H1=1，L1=0 时，左上角的 LED 发光。

如图 7-19 所示，采用 2 块 5×7 LED 点阵设计一个显示电路，显示内容为第 1s 显示数字 31，第 2s 显示数字 46，轮流显示。

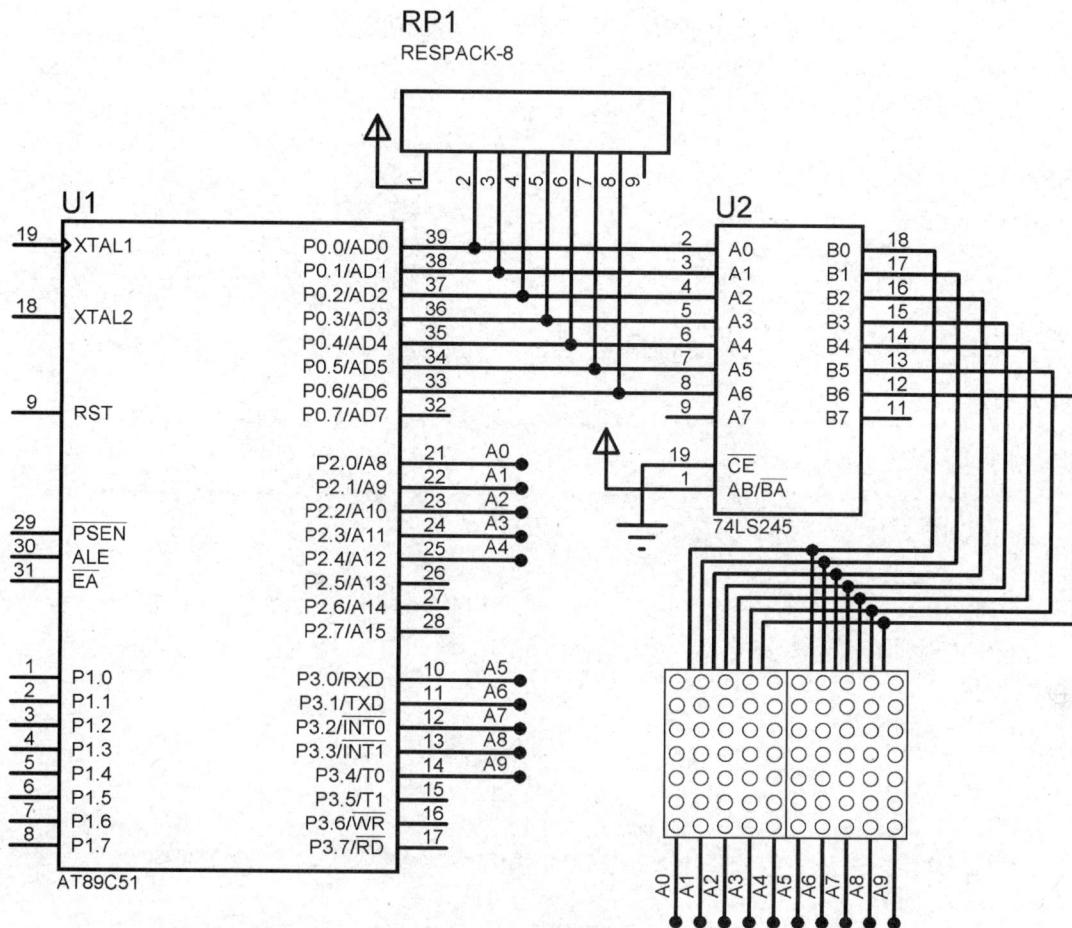

图 7-19 5×7 LED 点阵显示电路

参考程序如下：

```
/*******************************************************
程序名称：program7-4.c
程序功能：5×7 LED 点阵显示程序
*******************************************************/
#include<reg52.h>                              //加载头文件
#include<intrins.h>                            //加载头文件
/*******************************************************
数据类型定义
*******************************************************/
#define uchar unsigned char                    //定义无符号字符型
/*******************************************************
数组定义：5×7 LED 点阵显示数字 0～9 的编码表
*******************************************************/
uchar code LED5X7_CODE[]=
{
```

5×7 LED 点阵系统
参考程序

0x00,	0x3E,	0x3e,	0x3e,	0x00,	//数字 0 的编码
0xff,	0xff,	0x80,	0xff,	0xff,	//数字 1 的编码
0x30,	0x36,	0x36,	0x36,	0x06,	//数字 2 的编码
0x36,	0x36,	0x36,	0x36,	0x00,	//数字 3 的编码
0x07,	0x77,	0x77,	0x77,	0x00,	//数字 4 的编码
0x06,	0x36,	0x36,	0x36,	0x30,	//数字 5 的编码
0x00,	0x36,	0x36,	0x36,	0x30,	//数字 6 的编码
0x3f,	0x3f,	0x3f,	0x3f,	0x00,	//数字 7 的编码
0x00,	0x36,	0x36,	0x36,	0x00,	//数字 8 的编码
0x06,	0x36,	0x36,	0x36,	0x00	//数字 9 的编码

```c
};;
/********************************************************************
延时 1ms 函数，定时器 T0 产生
********************************************************************/
void delay1ms()
{
    TH0=0xfc;                        //12MHz，1ms 定时初始值
    TL0=0x18;
    TR0=1;                           //开启定时器
    while(TF0==0);                   //查询判断有无溢出
    TF0=0;                           //若溢出，则清零
    TR0=0;                           //关闭定时器
}
/********************************************************************
显示函数，占用 10ms
********************************************************************/
void display(uchar k)
{
    uchar i,j=0x01;
    P3=0x00;                         //关闭 P3 口扫描
    for(i=0;i<5;i++)
    {
        P2=j;                        //P2 口开始扫描
        P0=LED5X7_CODE[k/10*5+i];
        delay1ms();
        j=_crol_(j,1);
    }
    P2=0x00;                         //关闭 P2 口扫描
    j=0x01;
    for(i=0;i<5;i++)
    {
        P3=j;                        //P3 口开始扫描
        P0=LED5X7_CODE[k%10*5+i];
        delay1ms();
        j=_crol_(j,1);
```

```
        }
    }
/*****************************************************************
主函数
*****************************************************************/
void main()
{
    uchar i;
    TMOD=0x01;
    while(1)
    {
        for(i=100;i>0;i--)          //显示函数循环 100 次，实现延时 10ms×100=1s
            display(31);            //显示数字 31
        for(i=100;i>0;i--)          //显示函数循环 100 次，实现延时 10ms×100=1s
            display(46);            //显示数字 46
    }
}
```

习题

一、单选题

1．8×8 LED 点阵共有（　　）个引脚。

A．8   B．12   C．16   D．64

2．若单片机采用列扫描方式驱动 8×8 LED 点阵，则扫描时间为（　　）。

A．1ms   B．10ms   C．100ms   D．1s

3．16×32 点阵可以由（　　）块 8×8 LED 点阵构成。

A．4   B．8   C．16   D．32

二、多选题

1．下列关于 74HC595 的说法中正确的是（　　）。

A．74HC595 是一款 8 位串入并出的移位寄存器

B．74HC595 具有级联能力，可以通过多个芯片级联扩展输出位数

C．74HC595 的 16 引脚为 V_{CC}

D．74HC595 的 8 引脚为 GND

2．下列关于 74HC595 的功能说法中正确的是（　　）。

A．当 DS 为低电平，MR 为高电平，SHCP 为上升沿时，移位寄存器存储低电平

B．当 DS 为高电平，MR 为高电平，SHCP 为上升沿时，移位寄存器存储高电平

C．当 MR 为低电平时，移位寄存器处于清零状态

D．当 MR 为高电平，SHCP 为下降沿时，移位寄存器处于保持状态

项目八　串行接口系统的仿真设计

任务 8.1　双机通信系统的仿真设计

学习目标

【知识目标】

（1）了解并掌握串行通信的原理。

（2）了解并掌握 51 单片机串行通信的使用方法。

【技能目标】

（1）了解并掌握单片机仿真软件 Proteus 的使用方法。

（2）了解并掌握单片机编译软件 Keil C51 的使用方法。

（3）了解并掌握单片机程序下载的方法。

（4）了解并掌握单片机最小系统的组成。

（5）通过双机通信系统的仿真设计初步了解并掌握单片机项目的开发步骤。

【思政目标】

通过双机通信系统的仿真设计，结合当时的社会热点，如北斗导航系统的成功发射，激发学生的爱国热情和民族自豪感。

双机通信系统导学材料

什么是单片机串口微课视频

8.1.1　单片机的串行接口

计算机与外界的信息交换称为通信，通常具有并行和串行两种通信方法。

并行通信是指数据字节的各位同时发送，并通过并行接口实现。例如，MCS-51 系列单片机的 P0~P3 口就是并行接口。当 P1 口作为输出接口时，CPU 将一个数据写入 P1 口，数据在 P1 口上并行地同时输出到外设；当 P1 口作为输入接口时，对 P1 口执行一次读操作，在 P1 口上输入的 8 位数据同时被读出。

串行通信是指数据字节一位一位串行地顺序传输，并通过串行接口实现。串行接口进行数据传输的主要缺点是其传输速度比并行接口的慢，但它能节省传输线，特别是当数据位数很多和传输距离远时，这一优点更加突出。串行通信只需几根信号线就可完成数据的传输，同时必须依靠一定的通信协议（包括设备的选通、传输的启动、格式、结束）。

串行通信有两种基本的通信方式，即异步通信和同步通信，近年来又出现了 I^2C 总线串行通信。

1. 异步通信

异步通信时的字符由四部分组成：起始位（占 1 位）、字符代码数据位（占 5~8 位）、奇偶校验位（占 1 位，也可以没有）、停止位（占 1 或 2 位），如图 8-1 所示。

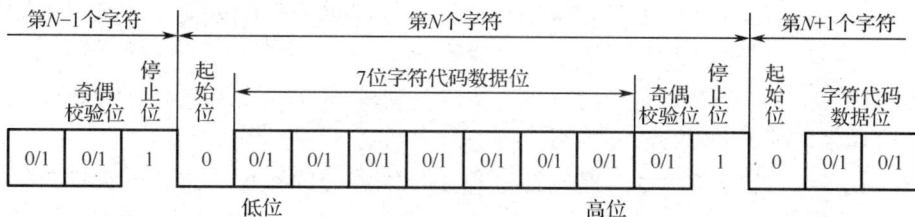

图 8-1 异步通信的格式

图 8-1 中给出的是 7 位字符代码数据位、1 位奇偶校验位和 1 位停止位，加上固定的 1 位起始位，共 10 位组成一个传送字符的格式。传送时数据的低位在前，高位在后；字符之间允许有不定长度的空闲位。起始位（0）作为联络信号，它用低电平告诉接收方开始传送，接下来的是字符代码数据位和奇偶校验位，停止位（1）标志一个字符的结束。

由图 8-1 可以看出，传送一个字符以起始位开始并以停止位结束。这就提供了区分和识别联络信号与数据信号的标志。传送开始前，收发双方要把所采用的信息格式（包括字符的字符代码数据位长度、停止位长度、有无奇偶校验位及采用的校验方式等）和数据传输速率即波特率做统一的约定。如果要改变信息格式和数据传输速率，则要求收发双方同时改变。

传送开始后，接收方不断检测传输线，看是否有起始位到来。当收到一系列的"1"（空闲位或停止位）之后，检测到一个"0"，说明起始位出现，就开始接收所规定的字符代码数据位、奇偶校验位及停止位。经过处理将停止位去掉，把字符代码数据位拼成一个并行字节，并且经校验无误才算正确地接收一个字符。一个字符接收完毕后，接收设备又继续测试传输线，监视"0"电平的到来（下一个字符开始），直到全部数据接收完毕。

由上述过程可以看到，异步通信是按字符传输的。每传输一个字符，就用起始位进行收发双方的同步。若接收设备和发送设备的时钟频率略有偏差，则不会因偏差的积累而导致错误。另外，字符之间的空闲位也能使这种偏差可能导致的错误得以减少。

2. 同步通信

对于同步通信，一帧数据包括由固定长度（如 100 个）的字符组成的一个数据块，其中每个字符由 5～8 位组成。在数据块的前面置有 1 或 2 个同步字符，最后是校验字符，如图 8-2 所示。在数据块中，字符与字符之间不允许留空。

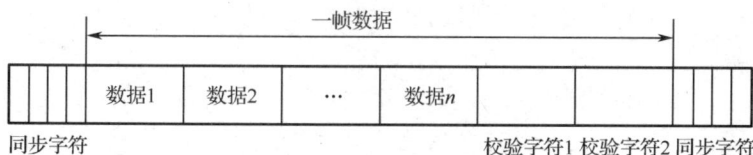

图 8-2 同步通信的格式

采用 2 个同步字符，称为双同步方式；采用 1 个同步字符，称为单同步方式。同步字符可以由用户来约定，也可以采用 ASCII 码中规定的同步空闲代码（SYN），即 16H。同步通信时，先发送同步字符，接收方检测到同步字符后，即准备接收数据，按约定的长度拼成一个个数据字节，直到整个数据接收完毕，经校验无误后，结束一帧数据的传输。

同步通信进行数据传输时,收发双方要保持完全的同步,因此要求收发设备必须使用同一时钟。在近距离通信时可以采用在传输中增加一根时钟信号线来解决;远距离通信时,可以采用锁相环技术,使接收方得到和发送方时钟频率完全相同的时钟信号。

同步通信的优点是有较高的传输速率(每秒可传输 56000bit 或更高),但硬件复杂。

3. 串行通信的制式

在串行通信中,通常在两个站之间通过单条传输线传输数据。根据数据的传输方向,串行通信可以分为三种制式:单工制式、半双工制式、全双工制式。串行通信的制式如图 8-3 所示。

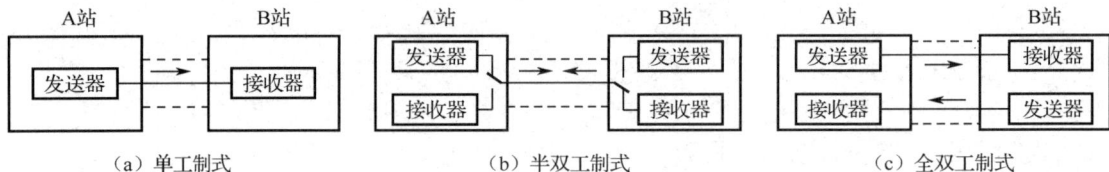

图 8-3 串行通信的制式

(1)单工制式。

单工制式指数据只能单方向传输。通信双方只具有发送数据或接收数据一种功能,一方(A 站)固定为发送端,另一方(B 站)固定为接收端,使用一条数据线。

单工制式一般用在只向一个方向传输数据的场合。例如,计算机与打印机之间的通信一般是单工制式。

(2)半双工制式。

半双工制式指使用一条数据线进行双向传输。A 站和 B 站都有发送器和接收器,但是发送和接收不能同时进行。当 A 站进行发送时,B 站只能进行接收;同理当 B 站进行发送时,A 站只能进行接收。对讲机就是一种采用半双工制式的通信设备。

(3)全双工制式。

全双工制式指通信双方能够同时进行发送和接收(使用两条数据线分别进行发送和接收),全双工制式的数据传输效率较高。51 单片机的串行接口就属于全双工制式。

4. 串行通信的校验

在串行通信过程中,往往要对数据传输的正确与否进行校验。校验是保证准确无误地传输数据的关键。常用的校验方法有奇偶校验、和校验、循环冗余校验。

(1)奇偶校验。

在发送数据时,数据位尾随的 1 位为奇偶校验位(1 或 0)。当约定为奇校验时,数据中"1"的个数与校验位中"1"的个数之和应为奇数;当约定为偶校验时,数据中"1"的个数与校验位中"1"的个数之和应为偶数。接收方与发送方的校验装置和方式应一致。接收数据时,对"1"的个数进行校验,若两者不一致,则说明传输过程中出现了差错。

(2)和校验。

所谓和校验是指发送方将所发数据块求和(或各字节异或),产生一个字节的校验字符(校验和)附加到数据块末尾。接收方接收数据的同时对数据块(除校验字节外)求和(或各字节异或),将所得的结果与发送方的校验和进行比较,若相符,则无差错,否则认为传输过程中出现了差错。

（3）循环冗余校验。

循环冗余校验是指对一个数据块校验一次，如对磁盘信息的访问、ROM 或 RAM 区的完整性等的校验，广泛应用于同步通信。

5. 传输速率与传输距离

（1）波特率。

波特率即数据传输速率，它表示每秒钟传送二进制代码的位数，其单位是 bit/s。波特率对 CPU 与外设的通信来说是一个很重要的参数。设数据传输速率是 240bit/s，而每个字符格式包含 10bit（1 个起始位、1 个停止位、8 个数据位），这时波特率为

$$10\text{bit} \times 240\text{s}^{-1} = 2400\text{bit/s}$$

每位码的传输时间 t_d 都为波特率的倒数，即

$$t_d = \frac{1\text{bit}}{2400\text{bit/s}} = 0.4165\text{ms}$$

波特率是衡量传输通信频宽的指标，它和数据传输速率并不一致。如上例中，因为除掉起始位和停止位，每个数据实际上只占 8bit，所以数据传输速率为

$$8\text{bit} \times 240\text{s}^{-1} = 1920\text{bit/s}$$

异步通信的传输速率为 50～19200bit/s，常用于计算机到终端机和打印机之间的通信、电报及无线电通信的数据发送等。

（2）传输速率与传输距离的关系。

串行接口或终端直接传输串行信息位流的最大传输距离（要求波形不发生畸变）与传输速率及传输线的电气特性有关。当传输线使用每 0.3m 有 50pF 电容的非平衡屏蔽双绞线时，传输距离随传输速率的增加而减小。当波特率超过 1000bit/s 时，最大传输距离迅速减小，如波特率为 9600bit/s 时最大传输距离减小到 76m。

6. 51 单片机串行接口的结构

MCS-51 系列单片机有一个可编程的全双工串行接口，它可作为通用异步收发器（UART），也可作为同步移位寄存器。其帧格式可为 8 位、10 位或 11 位，还可设置各种不同的波特率。通过 RXD（P3.0，串行数据接收端）和 TXD（P3.1，串行数据发送端）引脚与外界进行通信。UART 内部简化结构示意图如图 8-4 所示。

图 8-4 中有 2 个物理上独立的接收、发送缓冲器（SBUF），它们占用同一个地址 99H，可同时接收、发送数据。发送缓冲器只能写入，不能读出；接收缓冲器只能读出，不能写入。

串行发送和接收的速率与移位时钟同步，定时器 T1 作为串行通信的波特率发生

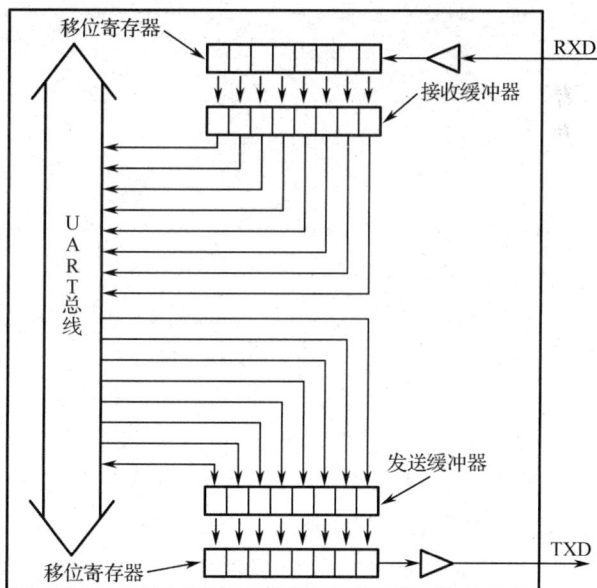

图 8-4 UART 内部简化结构示意图

器，T1 溢出率经 2 分频（或不分频）又经 16 分频作为串行发送或接收的移位时钟。移位时钟速率即波特率。

从图 8-4 中可以看出，接收器是双缓冲结构，在前一个字节从接收缓冲器读出之前，后一个字节开始被接收（串行输入至移位寄存器），但是在后一个字节接收完毕而前一个字节 CPU 未读取时，会丢失前一个字节的内容。串行接口的发送和接收都是以 SBUF 的名义进行读或写的，当向 SBUF 发送"写"命令时，即向发送缓冲器装载并开始由 TXD 引脚向外发送一帧数据，发送完毕后使发送中断标志位 T1=1。在串行接口接收中断标志位 RI（SCON.0）=0 的条件下，置串行接口允许接收位 REN（SCON.4）=1 就会启动接收，一帧数据进入输入移位寄存器，并装载到接收缓冲器中，同时使 RI=1。当执行读 SBUF 的命令时，由接收缓冲器取出信息通过内部总线发送至 CPU。

对于发送缓冲器，因为发送时 CPU 是主动的，所以不会产生重叠错误。

7. 串行控制寄存器 SCON

串行控制寄存器 SCON 是一个特殊功能寄存器，用于设定串行接口的工作方式、接收/发送控制及设置状态标志。其字节地址为 98H，可位寻址，各位的定义如表 8-1 所示。

表 8-1 SCON 各位的定义

位地址	9FH	9EH	9DH	9CH	9BH	9AH	99H	98H
位标志	SM0	SM1	SM2	REN	TB8	RB8	TI	RI

（1）串行接口工作方式选择位 SM0、SM1。

SM0 和 SM1（SCON.7 和 SCON.6）是串行接口工作方式选择位，可选择 4 种工作方式，如表 8-2 所示。

表 8-2 串行接口的工作方式

SM0	SM1	方式	说明	波特率
0	0	0	同步移位寄存器工作方式	$f_{osc}/12$
0	1	1	8 位数据的异步收发方式	可变
1	0	2	9 位数据的异步收发方式	$f_{osc}/64$ 或 $f_{osc}/32$
1	1	3	9 位数据的异步收发方式	可变

（2）多机通信控制位 SM2。

SM2（SCON.5）主要用于方式 2 和方式 3。若 SM2=1，则允许多机通信。多机通信协议规定，若第 9 位数据（D_8）为 1，则本帧为地址帧；若第 9 位数据为 0，则本帧为数据帧。当一个主机与多个从机通信时，所有从机的 SM2 都为 1。主机首先发一帧数据为地址，即某个从机号，其中第 9 位数据为 1，被寻址的某个从机收到数据后，将其中的第 9 位数据装入 RB8。从机依据收到的第 9 位数据（RB8 中）的值来决定从机可否再接收主机的数据，若 RB8=0，则说明是数据帧，使 RI=0，数据丢失；若 RB8=1，则说明是地址帧，将数据装入 SBUF 并使 RI=1，中断所有从机，被寻址的目标从机清除 SM2 以接收主机发来的一帧数据。其他从机的 SM2 仍保持 1。

若 SM2=0，则不属于多机通信，收到一帧数据后，无论第 9 位数据是 0 还是 1，都使 RI=1，收到的数据装入 SBUF。

在方式 1 下，若 SM2=1，则只有收到有效停止位时，才使 RI=1，以便接收下一帧数据。在方式 0 下，SM2 必须是 0。

（3）串行接口允许接收位 REN。

REN（SCON.4）是串行接口允许接收位。若 REN=1，则启动串行接口接收数据；若 REN=0，则禁止接收数据。

（4）发送第 9 位数据位 TB8。

TB8（SCON.3）主要用于方式 2 或方式 3，TB8 是发送第 9 位数据位，可以用软件规定其作用。TB8 可以用作数据的奇偶校验位，或者在多机通信中作为地址帧/数据帧的标志位。

在方式 0 和方式 1 下，TB8 未用。

（5）接收第 9 位数据位 RB8。

RB8（SCON.2）主要用于方式 2 或方式 3，RB8 是接收第 9 位数据位，可作为奇偶校验位或地址帧/数据帧的标志位。在方式 1 下，若 SM2=0，则 RB8 是收到的停止位。

（6）发送中断标志位 TI。

TI（SCON.1）为发送中断标志位。在方式 0 下，当串行发送第 8 位数据结束时，或者在其他方式下当串行发送停止位开始时，内部硬件使 TI=1，向 CPU 发送中断请求。在中断服务程序中，必须用软件使 TI=0，取消此中断请求。

（7）接收中断标志位 RI。

RI（SCON.0）为接收中断标志位。在方式 0 下，当串行接收第 8 位数据结束时，或者在其他方式下当串行接收停止位的中间时，内部硬件使 RI=1，向 CPU 发送中断请求。在中断服务程序中，必须用软件使 RI=0，取消此中断请求。

8. 电源控制寄存器 PCON

在电源控制寄存器 PCON 中，只有一位 SMOD 与串行接口工作有关，其格式如表 8-3 所示。

表 8-3　电源控制寄存器 PCON 的格式

位序	D7	D6	D5	D4	D3	D2	D1	D0
位符号	SMOD	—	—	—	GF1	GF0	PD	IDL

SMOD（PCON.7）为波特率倍增位。在串行接口的方式 1、方式 2、方式 3 下，波特率与 SMOD 有关，当 SMOD=1 时，波特率提高一倍。复位时，SMOD=0。

9. 波特率的计算

在串行通信中，收发双方对发送或接收的数据传输速率有一定的约定。通过软件可将单片机的串行接口编程为 4 种工作方式，其中方式 0 和方式 2 的波特率是固定的，而方式 1 和方式 3 的波特率是可变的，由定时器 T1 的溢出率决定。

串行接口的 4 种工作方式对应 3 种波特率。由于输入的移位时钟的来源不同，所以不同方式的波特率的计算公式也不同。

方式 0 的波特率的计算公式为 $f_{osc}/12$；方式 2 的波特率的计算公式为 $(2^{SMOD} \cdot f_{osc})/64$；方式 1、3 的波特率的计算公式为 $(2^{SMOD}/32) \cdot$（定时器 T1 的溢出率）。

当 T1 作为波特率发生器时，最典型的用法是使 T1 工作在自动再装入的 8 位定时器的方式（方式 2，且 TCON 的 TR1=1，以启动定时器）下，这时溢出率 n 取决于 TH1 中的计数值，即

$$n = \frac{f_{osc}}{12 \times (2^8 - TH1)}$$

使用单片机的串行接口时，常用的晶振频率为 11.0592MHz，所以选用的波特率也相对固定。在使用串行接口之前，应对它进行编程初始化，主要是设置产生波特率的定时器 T1、串行接口控制和中断控制。具体步骤如下。

（1）确定定时器 T1 的工作方式（编程 TMOD）。

（2）计算定时器 T1 的初始值，装载 TH1、TL1。

（3）启动定时器 T1（编程 TCON 中的 TR1）。

（4）确定串行接口控制（编程 SCON）。

（5）串行接口在中断方式下工作时，必须打开 CPU 的中断源（编程 IE、IP）。

8.1.2 双机通信系统的任务实施

【设计要求】

使用 AT89C2051 设计一个双机通信系统，电路图如图 8-5 所示，该系统包含甲、乙两机。系统开机后，甲机数码管显示数字"48"，乙机数码管显示数字"00"。当按下 K1 按键时，乙机数码管清零；当按下 K2 按键时，乙机数码管显示数字"99"；当按下 K3 按键时，甲机数码管显示值加 1；当按下 K4 按键时，甲机数码管显示值减 1。

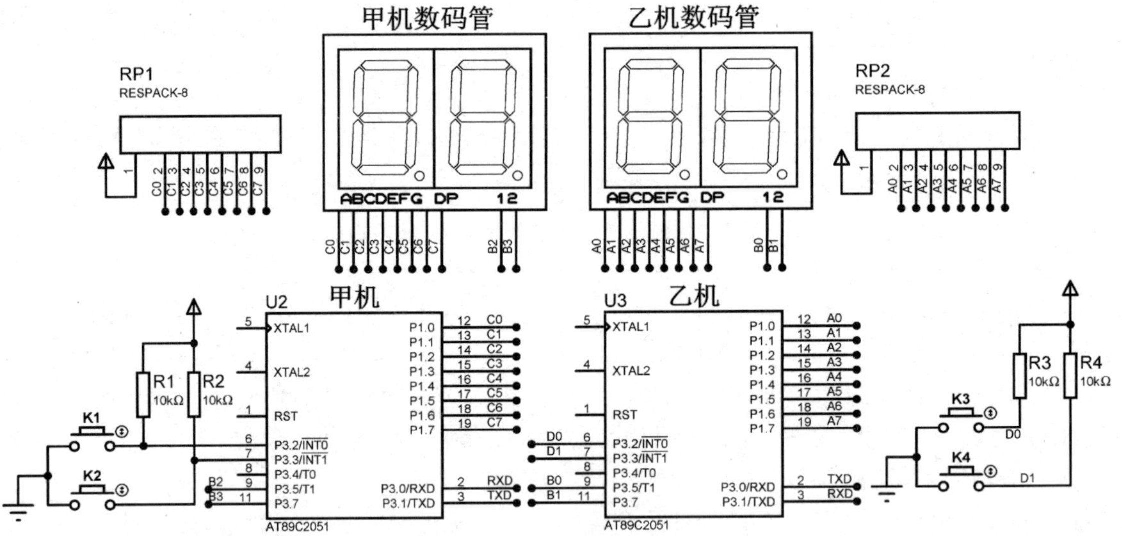

图 8-5 双机通信系统电路图

【任务分析】

根据设计要求，采用两位一体共阳极数码管作为双机通信系统的显示器件，K1、K2 按键分别与甲机的 P3.2、P3.3 口连接，K3、K4 按键分别与乙机的 P3.2、P3.3 口连接，两机采用串行通信，在连接时注意交叉，即甲机的 P3.0（RXD）口与乙机的 P3.1（TXD）口连接，甲机的 P3.1（TXD）口与乙机的 P3.0（RXD）口连接。

【实施步骤】

1. 添加元器件

打开 Proteus 仿真软件，按照表 8-4 添加元器件。

表 8-4　双机通信系统的元器件清单

元器件名称	所属类	所属子类
AT89C2051	Microprocessor ICs	8051 Family
RES	Resistors	Generic
7SEG-MPX2-CA-BLUE	Optoelectronics	7-Segment Displays
RESPACK-8	Resistors	Resistor Packs
BUTTON	Switches & Relays	Switches

2. 绘制仿真图

元器件全部添加后，在 Proteus ISIS 的原理图编辑窗口中按图 8-5 绘制双机通信系统仿真图。

3. 编写程序

单片机采用串行通信方式 1 进行双机通信，晶振频率为 11.0592MHz，波特率为 9600bit/s，因此需要将定时器 T1 作为波特率发生器，则 TMOD=0x20，TH1=TL1=0xfd。2 个按键分别接单片机的外部中断口，均采用下降沿触发方式进行按键检测。

在 Keil μVision5 中编写程序，实现双机通信系统的效果，甲机参考程序如下：

```
/*********************************************
程序名称：program8-1.c
程序功能：甲机程序
*********************************************/
#include<reg52.h>                    //加载头文件
/*********************************************
数据类型定义
*********************************************/
#define uchar unsigned char          //定义无符号字符型
/*********************************************
单片机引脚定义
*********************************************/
sbit P3_5=P3^5;
sbit P3_7=P3^7;
uchar TXD_DATA=0;                     //定义串行接口发送数据
uchar RXD_DATA=0;                     //串行接口接收数据存储单元
uchar DSP_DATA=58;                    //系统初始化显示值为58
/*********************************************
数组定义：共阳极数码管的编码表
*********************************************/
uchar code LED_7SEG_CA[]=
{  //0      1      2      3      4      5      6      7      8      9
```

甲机参考程序

```
        0xc0, 0xf9, 0xa4, 0xb0, 0x99, 0x92, 0x82, 0xf8, 0x80, 0x90
};
/**********************************************************************
数组定义：显示单元，2 字节
**********************************************************************/
uchar DISPLAY_DATA[]={0,0};
/**********************************************************************
延时 t ms 函数
**********************************************************************/
void delayms(uchar t)
{
    uchar i;
    while(t--)
        for(i=96;i>0;i--);
}
/**********************************************************************
显示函数
**********************************************************************/
void display()
{
    P3_7=0;
    P3_5=1;
    P1=LED_7SEG_CA[DISPLAY_DATA[0]];        //输出数据
    delayms(1);                             //延时 1ms
    P3_5=0;
    P3_7=1;
    P1=LED_7SEG_CA[DISPLAY_DATA[1]];        //输出数据
    delayms(1);                             //延时 1ms
}
/**********************************************************************
函数名：zd_int0()
功能描述：INT0 函数实现 K1 按键的识别
**********************************************************************/
void zd_int0() interrupt 0
{
    TXD_DATA=0xaa;                          //发送 0xaa 数据
    SBUF=TXD_DATA;                          //串行接口发送数据
}
/**********************************************************************
函数名：zd_int1()
功能描述：INT1 函数实现 K1 按键的识别
**********************************************************************/
void zd_int1() interrupt 2
{
    TXD_DATA=0xbb;                          //发送 0xbb 数据
```

```
        SBUF=TXD_DATA;                        //串行接口发送数据
}
/*************************************************************************
函数名：zd_rtx ()
功能描述：串行接口中断函数
*************************************************************************/
void zd_rtx () interrupt 4
{
    if(TI==1)
        TI=0;
    if(RI==1)
    {
        RI=0;                                 //清除接收中断标志位，表示接收完毕
        RXD_DATA=SBUF;                        //保存接收的数据
        if(RXD_DATA==0xaa)
        {
            DSP_DATA++;                       //显示值加 1
            if(DSP_DATA==100)                 //若大于 100，则清零
                DSP_DATA=0;
        }
        else if(RXD_DATA==0xbb)
        {
            DSP_DATA--;
            if(DSP_DATA==255)                 //若溢出，则置为最大值
                DSP_DATA=99;
        }
    }
}
/*************************************************************************
函数名：void main()
功能描述：主函数
*************************************************************************/
void main()
{
    SCON=0x50;                               //采用串行接口方式 1，允许接收，波特率可变
    TMOD=0x20;                               //T1 作为波特率发生器
    TH1=0xfd;                                //波特率为 9600bit/s，晶振频率为 11.0592MHz
    TL1=0xfd;
    EA=1;                                    //开启中断总开关
    EX0=1;                                   //开启 INT0 开关
    EX1=1;                                   //开启 INT1 开关
    ES=1;                                    //开启串行接口中断
    IT0=1;                                   //INT0 采用下降沿触发方式
    IT1=1;                                   //INT1 采用下降沿触发方式
    TR1=1;                                   //T1 开启工作
```

```
    while(1)
    {
        DISPLAY_DATA[0]=DSP_DATA/10;
        DISPLAY_DATA[1]=DSP_DATA%10;
        display();
    }
}
```

乙机参考程序如下:

```
/**********************************************************
程序名称: program8-2.c
程序功能: 乙机程序
**********************************************************/
#include<reg52.h>
#define uchar unsigned char
/**********************************************************
单片机引脚定义
**********************************************************/
sbit P3_5=P3^5;
sbit P3_7=P3^7;
bit FLAG_DISPLAY=0;
uchar TXD_DATA=0;                         //定义串行接口发送数据
uchar RXD_DATA=0;                         //串行接口接收数据存储单元
/**********************************************************
数组定义: 共阳极数码管的编码表
**********************************************************/
uchar code LED_7SEG_CA[]=
{    //0    1    2    3    4    5    6    7    8    9
    0xc0, 0xf9, 0xa4, 0xb0, 0x99, 0x92, 0x82, 0xf8, 0x80, 0x90
};
/**********************************************************
数组定义: 显示单元, 2 字节
**********************************************************/
uchar DISPLAY_DATA[]={0,0};
/**********************************************************
延时 t ms 函数
**********************************************************/
void delayms(uchar t)
{
    uchar i;
    while(t--)
        for(i=96;i>0;i--);
}
/**********************************************************
显示函数
```

```
******************************************************************/
void display()
{
    P3_7=0;
    P3_5=1;
    P1=LED_7SEG_CA[DISPLAY_DATA[0]];          //输出数据
    delayms(1);                               //延时 1ms
    P3_5=0;
    P3_7=1;
    P1=LED_7SEG_CA[DISPLAY_DATA[1]];          //输出数据
    delayms(1);                               //延时 1ms
}
/******************************************************************
函数名：zd_int0()
功能描述：INT0 函数实现 K1 按键的识别
******************************************************************/
void zd_int0() interrupt 0
{
    TXD_DATA=0xaa;                            //发送 0xaa 数据
    SBUF=TXD_DATA;                            //串行接口发送数据
}
/******************************************************************
函数名：zd_int1()
功能描述：INT1 函数实现 K1 按键的识别
******************************************************************/
void zd_int1() interrupt 2
{
    TXD_DATA=0xbb;                            //发送 0xbb 数据
    SBUF=TXD_DATA;                            //串行接口发送数据
}
/******************************************************************
函数名：zd_rtx ()
功能描述：串行接口中断函数
******************************************************************/
void zd_rtx () interrupt 4
{
    if(TI==1)
        TI=0;
    if(RI==1)
    {
        RI=0;                                 //清除接收中断标志位，表示接收完毕
        RXD_DATA=SBUF;                        //保存接收的数据
        if(RXD_DATA==0xaa)
            FLAG_DISPLAY=0;
        else if(RXD_DATA==0xbb)
```

```
                    FLAG_DISPLAY=1;
        }
}
/************************************************************************

函数名：void main()
功能描述：主函数
************************************************************************/
void main()
{
    SCON=0x50;                          //采用串行接口方式 1，允许接收，波特率可变
    TMOD=0x20;                          //T1 作为波特率发生器
    TH1=0xfd;                           //波特率为 9600bit/s，晶振频率为 11.0592MHz
    TL1=0xfd;
    EA=1;                               //开启中断总开关
    EX0=1;                              //开启 INT0 开关
    EX1=1;                              //开启 INT1 开关
    ES=1;                               //开启串行接口中断
    IT0=1;                              // INT0 采用下降沿触发方式
    IT1=1;                              // INT1 采用下降沿触发方式
    TR1=1;                              //T1 开启工作
    while(1)
    {
        if(FLAG_DISPLAY==0)
        {
            DISPLAY_DATA[0]=0;
            DISPLAY_DATA[1]=0;
        }
        else
        {
            DISPLAY_DATA[0]=9;
            DISPLAY_DATA[1]=9;
        }
        display();
    }
}
```

4. 系统仿真

当 Keil C51 编译成功后，会自动产生 HEX 文件，接着打开之前绘制的 Proteus 仿真图，双击 AT89C2051，弹出"Edit Component"对话框，单击"Program File"中的文件夹按钮，在弹出的"Select File Name"对话框中，选择之前编译生成的 HEX 文件，单击"打开"按钮，返回"Edit Component"对话框，单击"OK"按钮，即可装入 HEX 文件。

双机通信系统
仿真效果视频

接着单击 Proteus ISIS 编辑界面左下角的运行按钮 ▶，即可观察是否能够实现双机通信系统的显示效果，如图 8-6 所示。

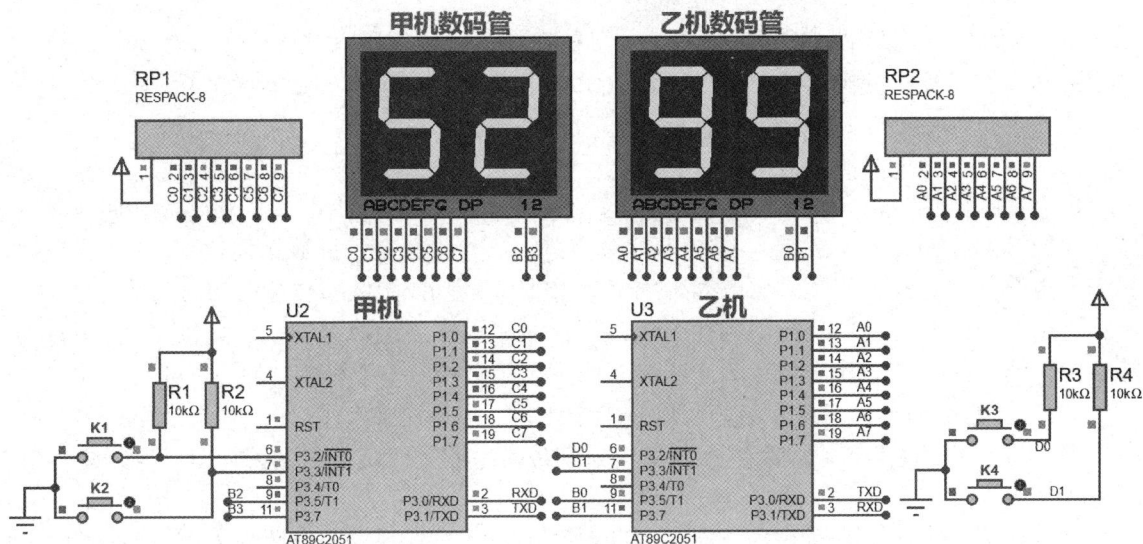

图 8-6　双机通信系统仿真效果图

任务 8.2　电子密码锁的仿真设计

✎ 学习目标

电子密码锁
导学材料

【知识目标】

（1）了解并掌握 I^{2}C 总线的基本原理。

（2）了解并掌握 E^{2}PROM 存储器 AT24C02 的使用方法。

【技能目标】

（1）了解并掌握单片机仿真软件 Proteus 的使用方法。

（2）了解并掌握单片机编译软件 Keil C51 的使用方法。

（3）通过电子密码锁的仿真设计进一步掌握单片机项目的开发步骤。

【思政目标】

在电子密码锁的仿真设计中，鼓励学生勇于尝试新技术、新方法，不断优化电子密码锁的设计和功能，培养学生的创新意识；引导学生认识到电子密码锁在保障公共安全、促进社会发展方面的重要作用，提升学生的社会责任感。

8.2.1　I^{2}C 通信

什么是 I^{2}C 通信
微课视频

1. I^{2}C 总线简介

I^{2}C 总线（Inter IC Bus）是 Philips 半导体公司推出的一种串行总线，是一种用于 IC 器件之间的二线制同步串行通信总线。它通过 SDA（串行数据线）及 SCL（串行时钟线）两根线与总线上的器件进行信息传输，并根据地址识别每个器件。I^{2}C 总线上允许连接多种接口电路，如 A/D 转换器及 D/A 转换器、实时时钟/日历、LCD 驱动器、键盘接口等，也可以连接串行 E^{2}PROM。

I²C 总线是由 SDA 和 SCL 构成的串行总线，可以发送和接收数据。在 CPU 和被控器件之间双向传输，最高传输速率为 400Kbit/s。SDA 用于地址、数据的输入和数据的输出，使用时需要加上拉电阻；SCL 为器件数据传输的同步时钟信号。

2. I²C 总线的特点

（1）二线传输，I²C 总线上的所有主器件（单片机、微处理器等）、外围器件等都连接到同名端的 SDA、SCL 上。按照二线制串行通信总线的传输规定，把数据传输到总线上的器件称为发送器，从总线上接收数据的器件称为接收器。可以控制数据传输的器件称为主器件，也称为主机，通常就是单片机或其他微处理器。主机需要启动数据的发送和接收，并提供传输时钟，这个系统上的其他芯片即从器件，也称为从机。

（2）I²C 总线是一种多主系统，可以有多个主机。当系统中有多个主机时，任何一个主机在 I²C 总线上工作时都可以成为主控制器，通常最常用的方式是单主系统，即只有一个主机，其余为从机。

（3）在 I²C 总线上传输时，主机或从机都可能处于发送或接收的工作方式，当然有少数从机，如显示芯片，只能处于接收方式。

（4）所有外围器件都可以采用器件地址和引脚地址的硬件编址方法，避免了片选线的连接方法。

（5）所有带 I²C 总线接口的外围器件都具有应答功能，当片内有多个单位地址时，数据读/写都有自动加 1 的功能。

（6）I²C 总线上的数据传输速率一般为 100Kbit/s，目前有的器件可以达到 400Kbit/s 以上，在高速模式下可达 3.4Mbit/s。连接到总线上的器件数仅受总线小于 400pF 的电容和器件地址容量的限制。

3. I²C 总线的组成及基本工作原理

I²C 总线只有两根双向信号线。一根是 SDA，另一根是 SCL。所有连接到 I²C 总线上的器件的数据线都接到 SDA 上，时钟线都接到 SCL 上。I²C 总线的基本架构如图 8-7 所示。

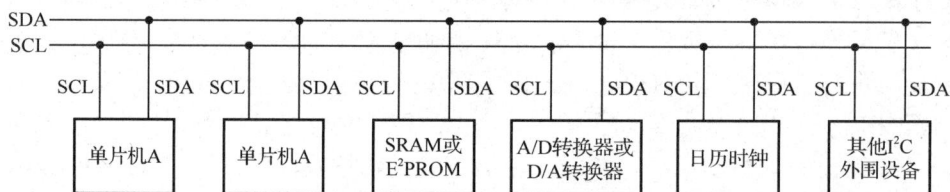

图 8-7　I²C 总线的基本架构

SDA 和 SCL 都是双向 I/O 接口线。接口电路为开漏输出，要通过上拉电阻接正电源。当总线空闲时，SDA 和 SCL 都是高电平。连接总线的外围器件都是 CMOS 器件，输出级也是开漏电路，在总线上消耗的电流很小。因此，总线上扩展的器件数主要由电容负载来决定。因为每个器件的总线接口都有一定的等效电容，而线路中的电容会影响总线的传输速率，当电容过大时，有可能造成传输出错，所以其负载能力定为 400pF。由此可估算出总线长度和所接器件数。

主机用于启动总线传输数据，并产生时钟以开放传输的器件，此时任何被寻址的器件均被认为是从机。在总线上，主和从、发送和接收的关系不是永久的，仅取决于此时数据传输的方向。

I²C 总线的数据传输一般按以下方式进行：如果主机要把信息传输至器件 1，则主机首先寻址器件 1，然后主机（主发送）把数据传输至器件 1（从接收），最后由主机终止传输；如果主机要从器件 1 中接收信息，则首先由主机寻址器件 1，然后主机（主接收）接收器件 1（从发送）的数据，最后由主机终止接收。在这种情况下，主机负责产生定时时钟和终止数据传输。

4. I²C 总线的传输时序

在 I²C 总线上，每位数据位的传输都与时钟脉冲相对应。逻辑 0 和逻辑 1 的信号电平取决于相应的电源电压。

在传输数据时，SCL 为高电平器件，SDA 上的数据必须保持稳定，只有在 SCL 为低电平期间，SDA 上的电平状态才允许变化。数据传输时序图如图 8-8 所示。

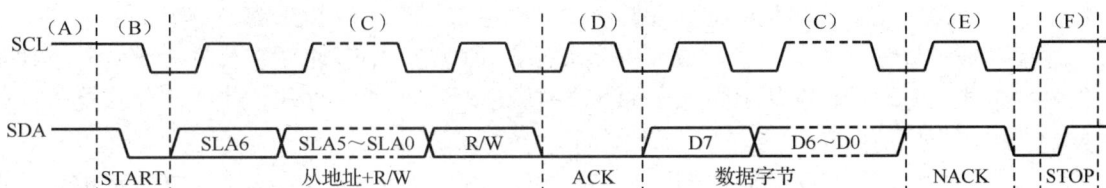

图 8-8　数据传输时序图

对于不同总线状态的说明如下。

（1）A 端：总线空闲阶段。

SDA 和 SCL 均为高电平，处于总线空闲状态。

（2）B 端：启动（START）数据传输阶段。

当 SCL 为高电平时，如果 SDA 从高电平下降为低电平，则表示处于启动状态。只有出现启动信号后，其他命令才有效。

（3）C 端：数据传输阶段。

在出现启动信号后，当 SCL 为高电平时，SDA 保持不变，这时 SDA 的状态就表示要传输到数据。当 SCL 为低电平时，SDA 上的数据位发生改变，每位数据位都需要 1 个时钟脉冲，一次顺序发送 1 字节的 8 位。

注意：所有数据和地址字节都是首先发送最高位即地址的第 6 位（但它是该字节的第 7 位），SLA5～SLA0 即地址的第 5～0 位，R/W 是读/写位，是该字节的第 0 位；图 8-8 中第 2 个 C 端传输的是数据，在 SDA 上串行传输数据时，最高有效位在前，最低有效位在后，与时钟同步，这个通信协议支持 8 位的双向数据传输。

（4）D 端：应答（ACK）信号阶段。

ACK 信号是主机对从机工作状态的一种检测。主机查询到从机有 ACK "0" 信号输出，说明其内部定时写周期结束，可以写入新的内容。每当从机接收完一个 8 位的写入地址或数据，就会在第 9 个时钟周期出现 ACK 信号，如图 8-8 中的 D 端所示。此时，从机发送一个 ACK "0" 信号。反之，当主机接收完来自从机的数据后，单片机也应向 SDA 发送 ACK 信号。单片机是通过发送起始状态及芯片地址做应答查询的。

（5）E 端：非应答（NACK）信号阶段。

当 SCL 为高电平时，如果 SDA 为高电平，则表示收到的是 NACK 信号。在典型的数据传输中，收到 NACK 信号表示所寻址的从机没有准备好或不在总线上。一个处于接收状态的主机发送 NACK 信号，表示这是数据传输的最后一个字节。

（6）F 端：停止（STOP）数据传输阶段。

当 SCL 为高电平时，如果 SDA 从低电平上升到高电平，则表示这是一个停止信号。在出现停止信号后，所有操作都停止。

每个数据的传输都由启动信号开始，到停止信号结束。在启动信号与停止信号之间传输的字节数由主机决定，理论上对字节数没有限制。

发送到 SDA 上的每个字节必须为 8 位，每次传输的字节数不限，每个字节后面必须跟 1 个应答位。传输数据时，先传输最高位。如果从机不能接收下一个字节（例如，正在处理一个内部中断，在这个中断处理完前不能接收 I²C 总线上的字节），则可以使时钟保持低电平，迫使主机处于等待状态。当从机准备好接收下一个字节后，释放 SCL 继续传输。

通常，被寻址的从机必须在收到每个字节后做出响应，若从机正在处理一个实时事件不能接收或不能认可地址，则从机必须使 SDA 保持高电平，此时，主机产生 1 个结束信号使传输异常结束。

I²C 总线发送信号的第 1 个字节用来确定主机所选择的从机。该字节的高 7 位组成了从机地址，最低位确定了信息的传输方向。当最低位为 0 时，表示主机把信息写到所选择的从机上；当最低位为 1 时，表示主机将由从机读取信息。在系统中，各器件将自己的地址与主机送到总线上的地址进行比较。如果匹配，则该器件为被主机寻址的从机，是接收还是发送由最低位确定。

绝大多数单片机应用系统中采用单主系统，即只有 1 个主机，其余为从机，在单主系统中，I²C 总线只需采用主方式工作，此时总线数据的传输状态要简单得多，没有总线竞争与同步的问题，只有单片机对接点器件的读/写操作。这样就可以采用不带 I²C 总线接口的单片机，利用其 I/O 接口实现对 I²C 总线上器件的读/写操作。

5. I²C 总线的寻址字节

在主机发出启动信号后要再传输一个寻址字节：7 位从机地址、1 位数据传输方向位（0 表示主机发送数据，1 表示主机接收数据），其格式如图 8-9 所示。

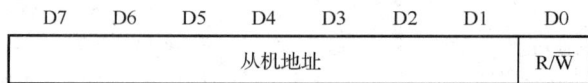

D7	D6	D5	D4	D3	D2	D1	D0
从机地址							R/$\overline{W}$

图 8-9 寻址字节的格式

D7～D1 位组成从机地址。D0 位是数据传输方向位。当主机发送地址时，总线上的每个从机都将这 7 位地址码与自己的地址进行比较。如果相同，则认为自己正在被主机寻址。从机地址由固定部分和可编程部分组成。

8.2.2 AT24C02 的使用

E²PROM 的优点是体积小、功耗低、占用 I/O 接口线少、性价比高。典型产品如 Atmel 公司的 AT24C02，其引脚排列图及与 51 单片机的连接如图 8-10 所示。

AT24C02 芯片的使用微课视频

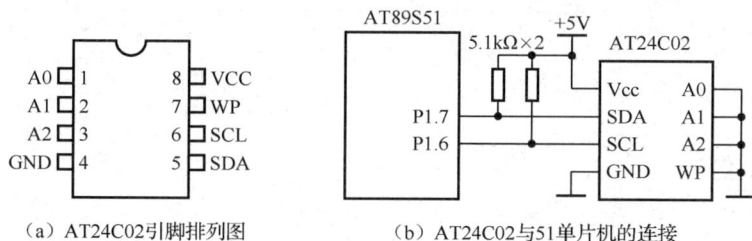

（a）AT24C02引脚排列图　　　　（b）AT24C02与51单片机的连接

图 8-10　AT24C02 的引脚排列图及与 51 单片机的连接

AT24C02 内含 256Byte（2KB），擦写次数大于 10000，写入时间短（小于 10ms）。在图 8-10 中，由于仅扩展一个器件，所以将 A2、A1、A0 这 3 条地址线接地，WP 为写保护控制端，接地时允许写入。SDA 为数据输入/输出线，SCL 为串行时钟线。

AT24C 系列单片机器件地址表如表 8-5 所示。

表 8-5　AT24C 系列单片机器件地址表

器件型号	字节容量/B	寻址字节				内部地址字节数	页面写字节数	最多可挂器件数
		固定标识	片选		R/W			
AT24C01A	128	1010	A2	A1	A0	1/0	1	8	8
AT24C02	256	1010	A2	A1	A0	1/0	1	8	8
AT24C04	512	1010	A2	A1	A0	1/0	1	16	4
AT24C08A	1K	1010	A2	A1	A0	1/0	1	16	2
AT24C16A	2K	1010	P2	P 1	P 0	1/0	1	16	1
AT24C32A	4K	1010	A2	A1	A0	1/0	2	32	8
AT24C64A	8K	1010	A2	A1	A0	1/0	2	32	8
AT24C128B	16K	1010	A2	A1	A0	1/0	2	64	8
AT24C256B	32K	1010	A2	A1	A0	1/0	2	64	8
AT24C512B	64K	1010	A2	A1	A0	1/0	2	128	8

由表 8-5 可见，AT24C02 地址的固定标识为 1010，器件引脚 A2、A1 和 A0 的不同连接可以选择 8 个同样的器件，片内 256B 可以由单字节寻址，页面写字节数为 8。

1. 写操作过程

对 AT24C02 写入时，单片机发出启动信号之后再发送的是控制字节，然后释放 SDA 并在 SCL 上产生第 9 个时钟信号。被选中的存储器在确认是自己的地址后，在 SDA 上产生一个 ACK 信号，单片机收到 ACK 信号后就可以传输数据了。

传输数据时，单片机首先发送一个字节的预写入存储单元的首地址，收到正确的 ACK 信号后，单片机逐个发送各字节，但每发送一个字节都要等待 ACK 信号。单片机发出停止信号后，启动 AT24C02 的内部写周期，完成数据写入工作（约 10ms 内结束）。

AT24C02 片内地址指针在收到每个字节后自动加 1，在芯片的一次装载字节数（页面字节数）限度内，只需输入首地址。当装载字节数超过一次装载字节数时，数据地址将"上卷"，前面的数据将被覆盖。

当要写入的数据传输完后，单片机应发出停止信号以结束写入操作。写入 n 个字节的数据格式如图 8-11 所示。

S	写控制字节	A	写入首地址	A	第1个字节	A	…	A	第*n*个字节	A	P

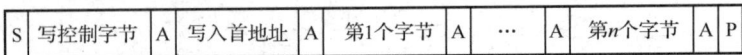

图 8-11　写入 *n* 个字节的数据格式

2. 读操作过程

对 AT24C02 读出时，单片机发送该器件的控制字节（"伪写"），发送完后释放 SDA 并在 SCL 上产生第 9 个时钟信号，被选中的存储器在确认是自己的地址后，在 SDA 上产生一个 ACK 信号作为响应。

然后，单片机发送一个字节的预读出存储单元的首地址，收到 ACK 信号后，单片机要重复一次启动信号并发出器件地址和读方向位（1），收到 ACK 信号后就可以读出字节，每读出一个字节，单片机都要回复 ACK 信号。当最后一个字节读完后，单片机应返回 NACK 信号（高电平），并发出停止信号以结束读出操作。

读出 *n* 个字节的数据格式如图 8-12 所示。

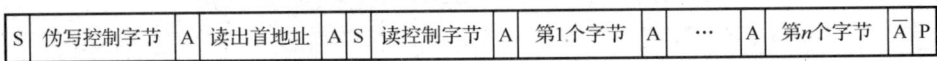

S	伪写控制字节	A	读出首地址	A	S	读控制字节	A	第1个字节	A	…	A	第*n*个字节	$\overline{A}$	P

图 8-12　读出 *n* 个字节的数据格式

8.2.3　电子密码锁的任务实施

【设计要求】

如图 8-13 所示，电子密码锁采用数码管进行显示，密码输入采用 4×3 矩阵式键盘。具体设计要求如下。

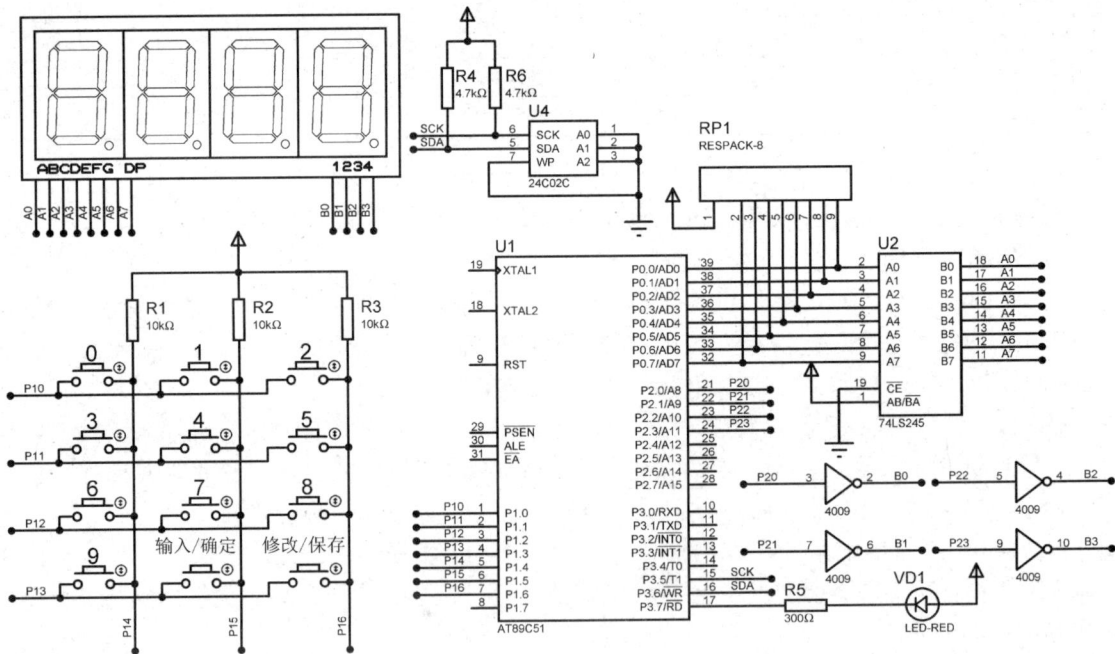

图 8-13　电子密码锁仿真图

（1）当系统开机时，数码管处于熄灭状态。

（2）当按下输入/确定按键后，数码管显示"—"，此时可通过数字按键输入密码，若密码输入完毕后，再次按下输入/确定按键，则系统自动判断该密码是否正确，若正确（初始密码为1234），则数码管黑屏，发光二极管点亮，表示开锁，5s后发光二极管自动熄灭；若错误，则数码管显示"Erro"，发光二极管不亮，此时需要重新按下输入/确定按键才可输入密码。

（3）在正确输入密码并开锁后，即发光二极管亮5s的时间内，可按下修改/保存按键进行密码修改，密码修改完后，可再次按下修改/保存按键，将新密码保存到 E^{2}PROM 中，其余时间不能修改密码。

【任务分析】

在本任务中，采用四位一体共阳极数码管进行显示。密码输入采用 4×3 矩阵式键盘，其中包括 0～9 十个数字按键，以及两个功能按键，每个功能按键都包含两种功能。E^{2}PROM 采用 AT24C02，用于存放四位数的密码，断电可保存数据，系统开机时读取 AT24C02 中的密码，修改密码后，将新密码保存到 AT24C02 中。

【实施步骤】

1. 添加元器件

打开 Proteus 仿真软件，按照表 8-6 添加元器件。注意：用 Proteus 仿真软件绘制单片机仿真图时，可以省略振荡电路和复位电路。

表 8-6 电子密码锁的元器件清单

元器件名称	所属类	所属子类
AT89C51	Microprocessor ICs	8051 Family
RES	Resistors	Generic
74LS245	TTL 74LS series	Transceivers
RESPACK-8	Resistors	Resistor Packs
7SEG-MPX4-CA	Optoelectronics	7-Segment Displays
LED-RED	Optoelectronics	LEDs
BUTTON	Switches & Relays	Switches
24C02C	Memory ICs	I2C Memories
4009	CMOS 4000 series	Buffers & Drivers

2. 绘制仿真图

元器件全部添加后，在 Proteus ISIS 的原理图编辑窗口中按图 8-13 绘制电子密码锁仿真图。

电子密码锁参考程序

3. 编写程序

在 Keil μVision5 中编写程序，实现电子密码锁的效果，参考程序如下：

```
/**********************************************************
程序名称：program8-3.c
程序功能：电子密码锁程序
**********************************************************/
```

```c
#include<reg52.h>                              //加载头文件
#include<intrins.h>                            //加载头文件
/**********************************************************************
数据类型定义
**********************************************************************/
#define uchar unsigned char                    //定义无符号字符型
/**********************************************************************
单片机引脚定义
**********************************************************************/
sbit LED=P3^7;                                 //开锁灯，灯亮表示锁打开，灯灭表示锁关闭
sbit IIC_SCK=P3^5;                             //I²C 器件的时钟端
sbit IIC_SDA=P3^6;                             //I²C 器件的数据端
/**********************************************************************
数组定义：共阳极数码管的编码表
**********************************************************************/
uchar code LED_7SEG_CA[]=
{   // 0     1     2     3     4     5     6     7     8     9
    0xc0, 0xf9, 0xa4, 0xb0, 0x99, 0x92, 0x82, 0xf8, 0x80, 0x90,
    //—    E     r     o     灭
    0xbf, 0x86, 0xaf, 0xa3, 0xff
};
/**********************************************************************
全局变量定义
**********************************************************************/
uchar KEY_NUMBER=0;                            //键盘存储单元号
uchar TIME=0;                                  //倒计时初始值
uchar IIC_ADDRESS=0x10;                        //I²C 总线存储首地址
uchar IIC_DATA;                                //I²C 总线读出的数据
uchar KEY_DATA[4]={0};                         //键盘临时存储单元
uchar MIMA[4]={1,2,3,4};                       //密码初始化
uchar IIC_BUFFER[4]={0};                       //片内 RAM 待发送或接收的数据存储单元
uchar DISPLAY_DATA[4]={0};                     //显示单元，4 字节
/**********************************************************************
延时 t ms 函数
**********************************************************************/
void delayms(uchar t)
{
    uchar i;
    while(t--)
        for(i=96;i>0;i--);
}
/**********************************************************************
延时 5μs 函数
**********************************************************************/
void delay5us()
```

```
{
    _nop_();
    _nop_();
    _nop_();
    _nop_();
}
```
/**
I²C 总线启动函数
**/
```
void iic_start()
{
    IIC_SDA=1;                          //启动 I²C 总线，将 SDA 置位
    IIC_SCL=1;                          //将 SCL 置位
    delay5μs();                         //延时 5μs
    IIC_SDA=0;                          //将 SDA 复位
    delay5μs();                         //延时 5μs
    IIC_SCK=0;                          //将 SCL 复位
}
```
/**
I²C 总线停止函数
**/
```
void iic_stop()
{
    IIC_SDA=0;                          //停止 I²C 总线，将 SDA 复位
    IIC_SCL=1;                          //将 SCL 置位
    delay5μs();                         //延时 5μs
    IIC_SDA=1;                          //将 SDA 置位
    delay5μs();                         //延时 5μs
    IIC_SDA=0;                          //将 SDA 复位
    IIC_SCL=0;                          //将 SCL 复位
}
```
/**
I²C 总线发送应答位函数
**/
```
void iic_mack()
{
    IIC_SDA=0;                          //将 SDA 复位
    IIC_SCL=1;                          //将 SCL 置位
    delay5μs();                         //延时 5μs
    IIC_SCL=0;                          //将 SCL 复位
    IIC_SDA=1;                          //将 SDA 置位
}
```
/**
I²C 总线发送非应答位函数
**/

```
void iic_mnack()
{
    IIC_SDA=1;                          //将 SDA 置位
    IIC_SCL=1;                          //将 SCL 置位
    delay5μs();                         //延时 5μs
    IIC_SCL=0;                          //将 SCL 复位
    IIC_SDA=0;                          //将 SDA 复位
}
```

/***
I²C 总线写字节函数
***/

```
void iic_write_byte(uchar iic_send)
{
    uchar i;
    for(i=0;i<8;i++)
    {
        iic_send<<=1;                   //发送数据左移，使发送位送入 CY 寄存器
        IIC_SDA=CY;                     //将 CY 寄存器的值赋值给 SDA
        _nop_();
        IIC_SCL=1;                      //将 SCL 置位
        delay5μs();                     //延时 5μs
        IIC_SCL=0;                      //将 SCL 复位
    }
    iic_mack();                         //I²C 总线发送应答位函数
}
```

/***
I²C 总线读字节函数
***/

```
void iic_read_byte()
{
    uchar i;
    for(i=0;i<8;i++)
    {
        IIC_SCL=1;                      //将 SCL 置位
        IIC_DATA<<=1;                   //接收数据左移
        IIC_DATA|=IIC_SDA;              //将 SDA 的值赋值给接收数据最低位
        IIC_SCL=0;                      //将 SCL 复位
    }
}
```

/***
I²C 总线读 n 个字节函数
***/

```
void iic_read_n_byte(uchar n)
{
    uchar i=0;
```

```c
        for(;n>0;n--)
        {
            iic_start();                    //启动 I²C 总线
            iic_write_byte(0xa1);           //发送读数据指令
            iic_read_byte();                //I²C 总线读字节
            IIC_BUFFER[i]=IIC_DATA;         //保存到片内 RAM
            iic_mnack();                    //发送非应答位
            iic_stop();                     //停止 I²C 总线
            i++;
        }
}
/*******************************************************************
I²C 总线写 n 个字节函数
*******************************************************************/
void iic_write_n_byte(uchar n)
{
    uchar i=0;
    iic_start();                        //启动 I²C 总线
    iic_write_byte(0xa0);               //发送写数据指令
    iic_write_byte(IIC_ADDRESS);        //发送 I²C 总线首地址
    for(;n>0;n--)
    {
        iic_write_byte(IIC_BUFFER[i]);  //发送 I²C 总线第 i 个数据
        i++;
    }
    iic_stop();                         //停止 I²C 总线
}
/*******************************************************************
AT24C02 写数据函数
*******************************************************************/
void write_24c02()
{
    IIC_ADDRESS=0x40;                   //I²C 总线存储首地址
    iic_write_n_byte(4);                //向 I²C 总线写入 4 个字节（不包括首地址）
}
/*******************************************************************
AT24C02 读数据函数
*******************************************************************/
void read_24c02()
{
    iic_write_n_byte(0);                //向 I²C 总线写入首地址
    iic_read_n_byte(4);                 //从 I²C 总线中读取 4 个字节
}
/*******************************************************************
显示函数
```

```
**********************************************************************/
void display()
{
    uchar i,j=0xfe;                              //共阴驱动
    for(i=0;i<4;i++)                             //共 4 个数码管
    {
        P2=j;                                    //点亮一个数码管
        P0=LED_7SEG_CA[DISPLAY_DATA[i]];         //输出数据
        j=_crol_(j,1);                           //准备点亮下一个数码管
        delayms(1);                              //延时 1ms
    }
}
/**********************************************************************
延时 20ms 函数
**********************************************************************/
void key_delay()
{
    uchar i;
    for(i=5;i>0;i--)
        display();
}
/**********************************************************************
按键查询函数
**********************************************************************/
void key_number(uchar i,j)
{
    while(P1==i)                                 //判断按键是否被释放
        display();                               //若未被释放,则等待
    if((FLAG_START==1)||(FLAG_XIUGAI==1))
    {
        KEY_DATA[KEY_NUMBER]=j;                  //存储按键值
        KEY_NUMBER++;                            //下一个数字按键值
        if(KEY_NUMBER==4)                        //判断数字按键是否溢出
            KEY_NUMBER=0;
    }
}
/**********************************************************************
密码输入/确定按键分支程序
**********************************************************************/
void key_start(uchar i)
{
    while(P1==i)                                 //判断按键是否被释放
        display();                               //若未被释放,则等待
    if(FLAG_START==0)
    {
```

```
            FLAG_START=1;                          //将密码输入标志位置位，允许输入密码
            KEY_DATA[0]=10;
            KEY_DATA[1]=10;
            KEY_DATA[2]=10;
            KEY_DATA[3]=10;
            KEY_NUMBER=0;                          //键盘存储单元号
        }
        else
        {
            FLAG_START=0;
            if((KEY_DATA[0]==MIMA[0])&&(KEY_DATA[1]==MIMA[1])&&(KEY_DATA[2]==
            MIMA[2])&&(KEY_DATA[3]==MIMA[3]))
            {
                KEY_DATA[0]=14;                    //若密码正确，则黑屏
                KEY_DATA[1]=14;                    //若密码正确，则黑屏
                KEY_DATA[2]=14;                    //若密码正确，则黑屏
                KEY_DATA[3]=14;                    //若密码正确，则黑屏
                FLAG_OK=1;                         //密码正确标志位置位
                TIME=5;                            //倒计时赋值 5s
                LED=0;                             //灯亮
            }
            else
            {
                KEY_DATA[0]=11;                    //若密码不正确，则显示"E"
                KEY_DATA[1]=12;                    //若密码不正确，则显示"r"
                KEY_DATA[2]=12;                    //若密码不正确，则显示"r"
                KEY_DATA[3]=13;                    //若密码不正确，则显示"o"
                FLAG_OK=0;                         //密码正确标志位复位
                LED=1;                             //灯灭
            }
        }
}
/*****************************************************************************
密码修改/密码保存按键分支程序
*****************************************************************************/
void key_xiugai(uchar i)
{
    uchar j;
    while(P1==i)                                  //判断按键是否被释放
        display();                                //若未被释放，则等待
    if(FLAG_XIUGAI==0)
    {
        if(FLAG_OK==1)
        {
            FLAG_XIUGAI=1;                         //将密码修改标志位置位，允许修改
```

```
                KEY_DATA[0]=10;
                KEY_DATA[1]=10;
                KEY_DATA[2]=10;
                KEY_DATA[3]=10;
            }
        }
        else
        {
            FLAG_XIUGAI=0;
            for(j=0;j<4;j++)                        //修改密码
            {
                MIMA[j]=KEY_DATA[j];
                IIC_BUFFER[j]=KEY_DATA[j];
                KEY_DATA[j]=14;                     //黑屏
            }
            write_24c02();                          //将新密码保存到 AT24C02 中
        }
}
/**********************************************************************
矩阵式键盘查询函数
**********************************************************************/
void key_scan()
{
    P1=0xf0;
    if(P1!=0xf0)                                    //判断 P1 口是否有按键被按下
    {
        key_delay();                               //延时 20ms 去抖动
        if(P1!=0xf0)                               //再次判断 P1 口是否有按键被按下
        {
            P1=0xfe;                               //查询第 1 行键盘
            switch(P1)
            {
                case 0xee: key_number(0xee,0);break; //数字 0 按键分支程序
                case 0xde: key_number(0xde,1);break; //数字 1 按键分支程序
                case 0xbe: key_number(0xbe,2);break; //数字 2 按键分支程序
            }
            P1=0xfd;                               //查询第 2 行键盘
            switch(P1)
            {
                case 0xed: key_number(0xed,3);break; //数字 3 按键分支程序
                case 0xdd: key_number(0xdd,4);break; //数字 4 按键分支程序
                case 0xbd: key_number(0xbd,5);break; //数字 5 按键分支程序
            }
            P1=0xfb;                               //查询第 3 行键盘
            switch(P1)
```

```
            {
                case 0xeb: key_number(0xeb,6);break;    //数字 6 按键分支程序
                case 0xdb: key_number(0xdb,7);break;    //数字 7 按键分支程序
                case 0xbb: key_number(0xbb,8);break;    //数字 8 按键分支程序
            }
            P1=0xf7;                                    //查询第 4 行键盘
            switch(P1)
            {
                case 0xe7: key_number(0xe7,9);break;    //数字 9 按键分支程序
                case 0xd7: key_start(0xd7);break;       //密码输入/确定按键分支程序
                case 0xb7: key_xiugai(0xb7);break;      //密码修改/密码保存按键分支程序
            }
        }
    }
}
/***********************************************************************
主函数
***********************************************************************/
void main()
{
    uchar i, j;
    KEY_DATA[0]=14;
    KEY_DATA[1]=14;
    KEY_DATA[2]=14;
    KEY_DATA[3]=14;
    IIC_ADDRESS=0x40;                   //I²C 总线的读数据首地址
    read_24c02();                       //读 AT24C02 存储的原始密码
    for(i=0;i<4;i++)
        MIMA[i]=IIC_BUFFER[i];
    while(1)
    {
        for(i=0;i<25;i++)               //4ms×10×25=1s，实现 1s

        {
            for(j=0;j<10;j++)           //4ms×10=40ms，每 40ms 查询一次键盘
            {
                DISPLAY_DATA[0]=KEY_DATA[0];
                DISPLAY_DATA[1]=KEY_DATA[1];
                DISPLAY_DATA[2]=KEY_DATA[2];
                DISPLAY_DATA[3]=KEY_DATA[3];
                display();
            }
            key_scan();
        }
        if(FLAG_OK==1)                  //若密码正确，则倒计时 5s
```

```
    {
        LED=0;                          //灯亮
        TIME--;                         //倒计时
        if(TIME==0)                     //判断倒计时是否结束
        {
            FLAG_OK=0;                  //密码正确标志位复位
            LED=1;                      //灯灭
        }
    }
    else
        LED=1;                          //灯灭
    }
}
```

4. 系统仿真

当 Keil C51 编译成功后，会自动产生 HEX 文件，接着打开之前绘制的 Proteus 仿真图，双击 AT89C51，弹出"Edit Component"对话框，单击"Program File"中的文件夹按钮，在弹出的"Select File Name"对话框中，选择之前编译生成的 HEX 文件，单击"打开"按钮，返回"Edit Component"对话框，单击"OK"按钮，即可装入 HEX 文件。

电子密码锁仿真效果视频

接着单击 Proteus ISIS 编辑界面左下角的运行按钮 ▶，即可观察是否能够实现电子密码锁的功能，如图 8-14 所示。

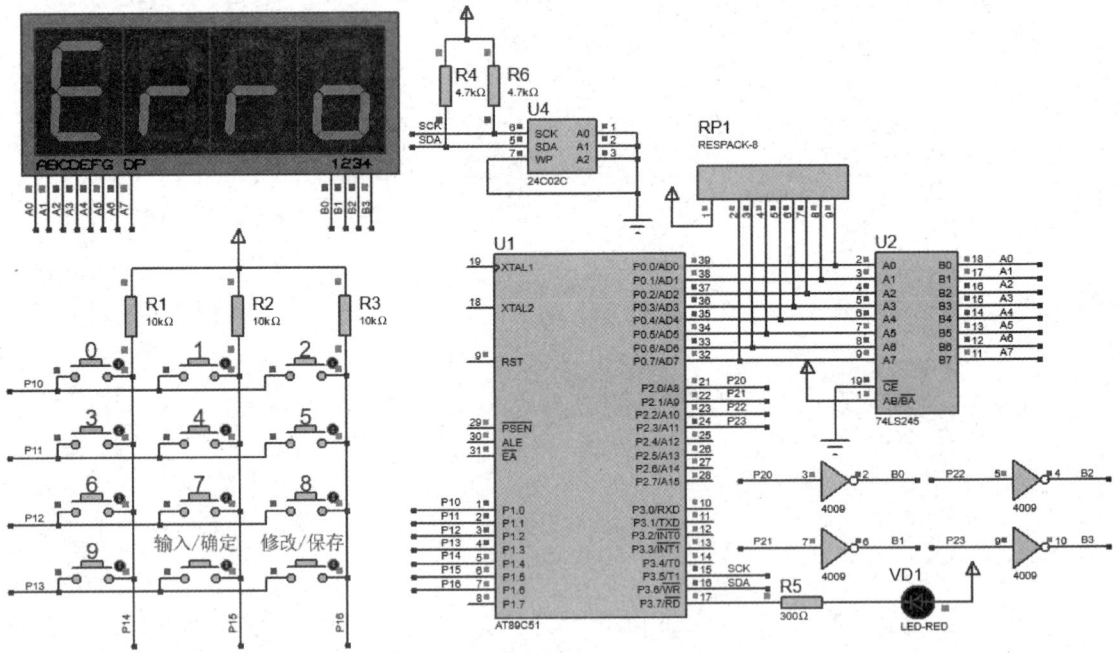

图 8-14　电子密码锁仿真效果图

素养小课堂

C51 语言与 C 语言的区别

C51 语言是 8051 单片机应用开发中最常使用的程序设计语言，它在 C 语言的基础上，针对 8051 单片机的特点进行了扩展，能直接对 8051 单片机进行操作，既有高级语言易读、开发效率高的优点，又有低级语言执行效率高的优点，已然成为最适合 51 单片机开发的实用高级语言。

C51 语言在语法规范、程序结构与设计方法上都与 C 语言基本相同，但在库函数、数据类型、变量存储模式等方面与 C 语言存在一些差别。

（1）库函数。C 语言的库函数是按通用微型计算机来定义的，C51 语言的库函数是按照 8051 单片机的特点来定义的。C51 语言有丰富的可直接调用的库函数，灵活使用库函数可使程序代码简单、结构清晰，并且易于调试和维护。每个库函数都在相应的头文件中给出了函数原型声明，用户如果需要使用库函数，就必须在源程序的开始处用预处理命令 "#include" 将有关的头文件包含进来。

（2）数据类型。针对 8051 单片机的特点，C51 语言在 C 语言的基础上增加了 4 种数据类型，它们分别是 bit、sfr、sfr16 和 sbit。

（3）变量存储模式。C 语言最初是为通用计算机设计的，在通用计算机中，只有一个程序和数据统一寻址的内存空间，而 C51 语言的变量存储模式与 8051 单片机的各种存储器紧密相关。

（4）数据存储类型。8051 单片机的存储区可分为片内 RAM、片外 RAM 和片内 ROM。

（5）C 语言没有处理单片机中断的定义，而 C51 语言中有专门的中断函数。

虽然 C 语言对语法的限制不太严格，用户在编写程序时有较大的自由度，但它毕竟是一种程序设计语言，与其他计算机程序设计语言一样，在采用 C 语言进行程序设计时，需要遵循一定的语法规则。

任何程序设计都离不开数据处理，一个程序如果没有数据，它就无法工作。数据在计算机内存中的存储情况由数据结构决定，C 语言的数据结构是以数据类型出现的，数据类型可分为基本数据类型和复杂数据类型，复杂数据类型由基本数据类型构造而成。C 语言中的基本数据类型有 char、int、short、long、float 和 double，对于 C51 编译器，short 与 int 相同，double 与 float 相同。

课后任务

1. 设计一个双机串口通信电路，电路图如图 8-15 所示，设计要求如下：当系统开机时，甲机驱动 8×8 LED 点阵显示汉字"丽"，乙机实现八路流水灯从上到下的流水效果，当按下 K1 按键后，八路流水灯实现从上到下和从下到上的流水效果的切换，当按下 K2 按键后，8×8 LED 点阵实现汉字"丽"和"水"的切换。波特率为 9600bit/s，晶振频率为 11.0592MHz。显示效果参见二维码。

课后任务 1
仿真效果视频

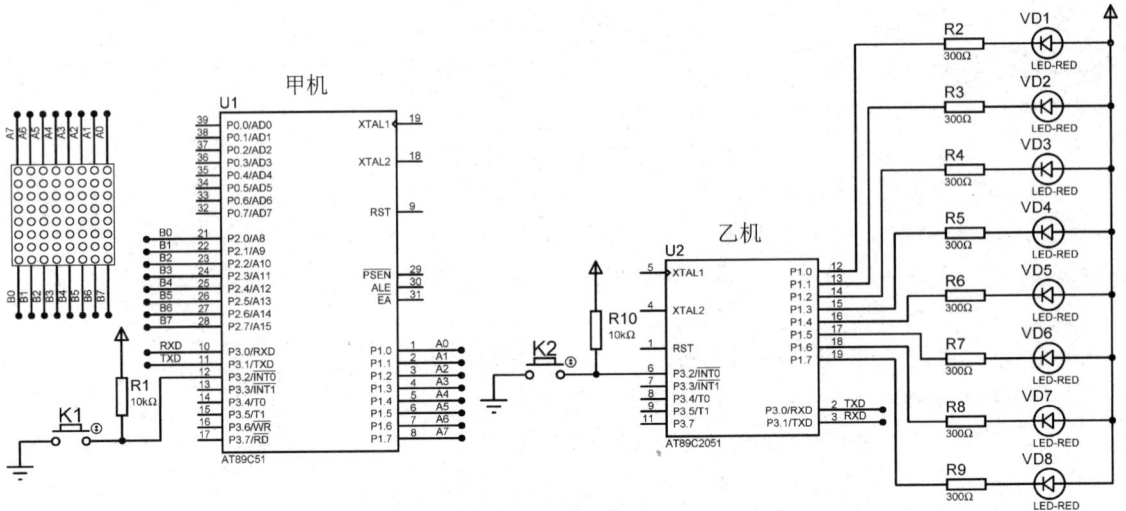

图 8-15　课后任务 1 电路图

2．在任务 8.2 的基础上添加功能，如图 8-16 所示。设计要求如下。

（1）系统开机后数码管显示学号后四位，在未处于密码输入状态时，按下学号按键，数码管也显示学号后四位。

（2）当按下密码输入按键后，数码管显示"—"，此时可通过数字按键输入密码，在密码输入状态时，若按下清除按键，则可将前一个数字清除，该位数码管重新显示"—"，若密码输入完毕后，按下确定按键，则系统自动判断该密码是否正确，若正确（初始密码为 1234），则数码管黑屏，发光二极管点亮，表示开锁，5s 后发光二极管自动熄灭；若错误，则数码管显示"Erro"，发光二极管不亮，此时需要重新按下密码输入按键才可输入密码。显示效果参见二维码。

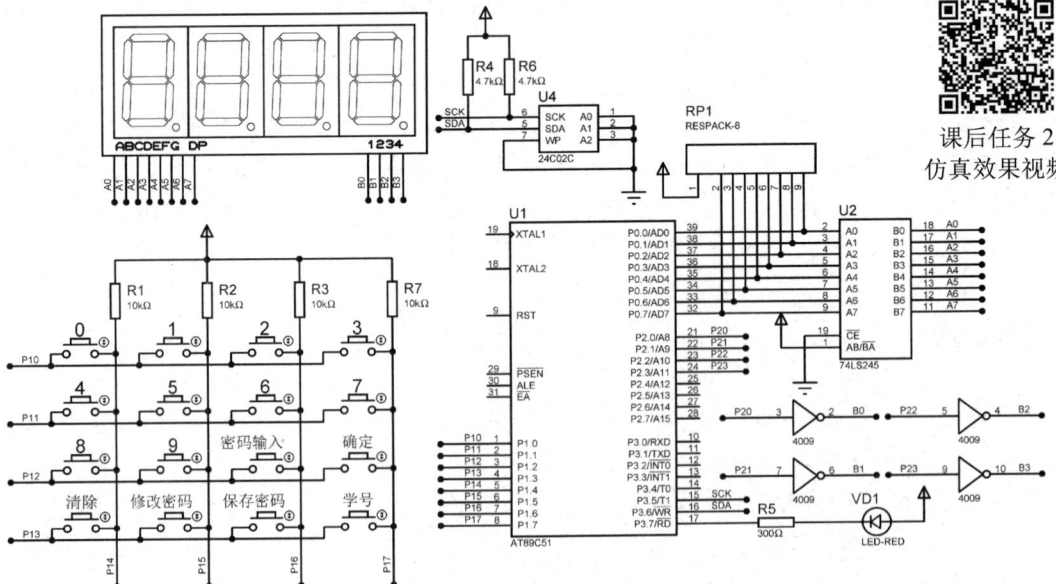

课后任务 2
仿真效果视频

图 8-16　课后任务 2 电路图

（3）密码输入模块采用 4×4 矩阵式键盘，其中 10 个按键为数字按键 0～9。

（4）在正确输入密码并开锁后，即发光二极管亮 5s 的时间内，可按下修改密码按键进行密码修改，按下保存密码按键后，保存新密码，并将新密码保存到 E²PROM 中，其余时间不能修改密码。

知识拓展 串口小工具

1. 波特率计算器

波特率计算器是一款用来计算单片机晶振频率和波特率的工具，一键即可计算并得到结果，如图 8-17 所示。用户可在"晶振频率"和"波特率"下拉列表中选择不同的晶振频率和波特率，单击"计算"按钮就可以计算出 T1 和 T2 的初始值及误差，非常简单实用。

2. UartAssist 串口调试助手

UartAssist 串口调试助手是在实际工程应用中，根据实际的普遍需求而开发的一款串口调试工具，具有很强的实用性，如图 8-18 所示，其主要具有以下特点。

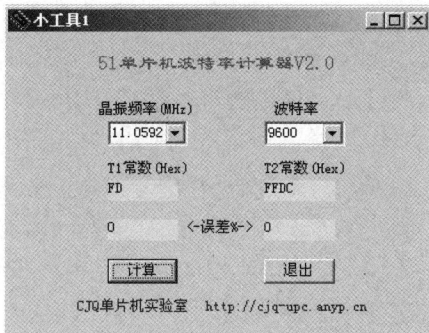

图 8-17 波特率计算器

图 8-18 UartAssist 串口调试助手

（1）支持任意波特率、校验位、数据位和停止位等。

（2）支持 ASCII/Hex 发送，发送和接收的数据可以在十六进制和 ASCII 码之间任意转换。

（3）可以自动在发送的数据末尾增加校验位，支持多种校验格式。

（4）支持间隔发送、循环发送、批处理发送，输入数据可以从外部文件导入，并且能够自动将收到的数据保存到磁盘文件。

（5）支持中/英文菜单，自动切换系统语言。

习题

一、单选题

1．单片机的串行接口工作在方式 0 时，RXD 引脚作为（　　　）引脚使用。

A．输入　　　　　　　B．输出　　　　　　　C．输入/输出　　　　　D．时钟

2．当串行接口采用中断方式工作时，发送或接收一帧数据后，其中断标志（　　　）。

A．会自动复位　　　B．需要软件复位　　　C．需要硬件复位　　　D．不允许操作

3．当 SCON=0x90 时，串行接口的工作状态为（　　　）。

A．工作在方式 2，允许接收　　　　　　　B．工作在方式 2，禁止接收

C．工作在方式 1，允许接收　　　　　　　D．工作在方式 3，禁止接收

4．单片机的晶振频率为 11.0592MHz，当 SCON=0x60，PCON=0x80，TH1=0xfa，TL1=0xfa 时，串行接口的波特率为（　　　）。

A．2.4Kbit/s　　　　B．4.8Kbit/s　　　　C．9.6Kbit/s　　　　D．19.2Kbit/s

5．控制串行接口工作方式的寄存器是（　　　）。

A．TCON　　　　　　B．PCON　　　　　　C．SCON　　　　　　D．TMOD

二、多选题

1．下列关于 80C51 单片机串行接口数据缓冲器的描述中正确的是（　　　）。

A．串行接口中有 2 个数据缓冲器

B．2 个数据缓冲器在物理上是相互独立的，具有不同的地址

C．发送缓冲器只能写入数据，不能读出数据

D．接收缓冲器只能读出数据，不能发送数据

2．下列关于 80C51 单片机串行接口内部结构的描述中正确的是（　　　）。

A．内部有一个可编程的全双工串行接口

B．串行接口可以作为通用异步接收/发送器，也可以作为同步移位寄存器

C．串行接口中设有 SCON

D．通过设置串行接口通信的波特率可以改变串行接口通信速率

3．下列关于 80C51 单片机串行接口发送控制器的作用描述中正确的是（　　　）。

A．作用一是将待发送的并行数据转为串行数据

B．作用二是在串行数据上自动添加起始位、可编程位和停止位

C．作用三是在数据转换结束后使 TI 自动置位

D．作用四是在中断被响应后使 TI 自动复位

4. 下列关于 80C51 单片机串行接口接收控制器的作用描述中正确的是（　　）。

A. 作用一是将来自 RXD 引脚的串行数据转为并行数据

B. 作用二是自动过滤掉串行数据中的起始位、可编程位和停止位

C. 作用三是在接收完成后使 RI 自动置位

D. 作用四是在中断被响应后使 RI 自动复位

5. 与方式 0 相比，方式 1 发生的下列变化中正确的是（　　）。

A. 通信时钟波特率是可变的，可由软件设置为不同速率

B. 数据帧由 11 位组成，包括 1 位起始位+8 位数据位+1 位校验位+1 位停止位

C. 发送数据由 TXD 引脚输出，接收数据由 RXD 引脚输入

D. 方式 1 可实现异步通信，而方式 0 只能实现串并转换

项目九 数字电压表的仿真设计

任务 9.1 ADC0809 数字电压表的仿真设计

学习目标

【知识目标】

（1）了解并掌握 A/D 转换的原理。

（2）了解并掌握 ADC0809 的使用方法。

【技能目标】

（1）了解并掌握单片机仿真软件 Proteus 的使用方法。

（2）了解并掌握单片机编译软件 Keil C51 的使用方法。

（3）了解并掌握单片机程序下载的方法。

（4）了解并掌握单片机最小系统的组成。

（5）通过 ADC0809 数字电压表的仿真设计初步了解并掌握单片机项目的开发步骤。

ADC0809 数字
电压表导学材料

【思政目标】

通过 ADC0809 数字电压表的仿真设计，强调精益求精、严谨细致的工作态度。要求学生通过反复测试、调试和改进，不断提高测量精度和降低误差。

什么是 A/D 转换
微课视频

9.1.1 A/D 转换

A/D 转换的作用是把一个模拟量转换为计算机能接收的数字量。模拟量是时间、数值都连续变化的物理量，如温度、压力、流量等，与此对应的电信号即模拟电信号。显然，模拟量要想输入计算机，首先要转换为数字量，才能被计算机接收。实现 A/D 转换的设备称为 A/D 转换器或 ADC。

1. A/D 转换的原理

A/D 转换器的种类很多，根据其转换原理可以分为逐次逼近式、双积分式、并行式、跟踪比较式和串并式等。目前，使用较多的是逐次逼近式和双积分式。逐次逼近式 A/D 转换器在精度、速度和价格方面都适中，是目前最常用的 A/D 转换器；双积分式 A/D 转换器具有精度高、抗干扰性好、价格低廉等优点，但速度较慢，经常应用在对速度要求不高的仪器仪表中。

这里主要讲述逐次逼近式 A/D 转换器的原理。逐次逼近式 A/D 转换器的原理即"逐位比较"，其过程类似用砝码在天平上称物体质量。具体方法是用一个二进制数作为计量单位的整数倍，并略去小于计量单位的部分，这样得到的整数量即数字量。显然，计量单位越小，量化误差就越小。图 9-1 所示为一个 N 位逐次逼近式 A/D 转换器的原理结构图，其由 N 位寄存器、D/A 转换器、比较器和时序与逻辑控制电路等部分组成，其中 N 位寄存器用于 N 位二进制数码。

当将模拟量 U_X 送入比较器后，启动信号通过时序与逻辑控制电路启动 A/D 开始转换。首先，N 位寄存器最高位（D_{n-1}）置位，其余位复位（相当于先放一个最重的砝码）。N 位寄存器的内容经 D/A 转换后，得到整个量程一半的模拟电压 U_N，然后与输入电压 U_X 进行比较。若 $U_X \geqslant U_N$，则保留 $D_{n-1}=1$；若 $U_X < U_N$，则 D_{n-1} 复位。然后，时序与控制逻辑电路使寄存器下一位（D_{n-2}）置位，与上次的结果一起经 D/A 转换后再与 U_X 进行比较。不断重复上述过程，直至判别出 D_0 取 1 还是取 0。此时，时序与逻辑控制电路发出转换结束信号（EOC）。这样经过 N 次比较后，N 位寄存器的内容就是转换后的数字量，经三态输出锁存器读出。转换过程就是逐次比较、逼近的过程。图 9-1 中的数字量可以采用并行方式输出，也可以采用串行方式输出。

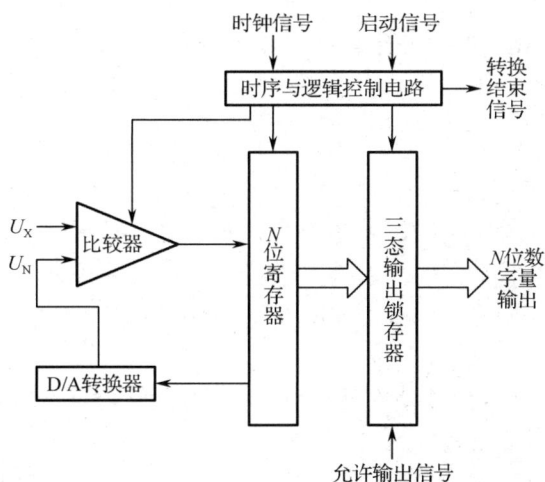

图 9-1　一个 N 位逐次逼近式 A/D 转换器的原理结构图

2. A/D 转换的主要技术指标

A/D 转换过程主要包括采样、量化与编码。采样时使模拟信号在时间上离散化；量化就是用一个基本的计量单位（量化电平）使模拟量变为一个整数的数字量；编码是把已经量化的模拟量（是量化电平的整数倍）用二进制数码、BCD 码或其他数码来表示。总之，量化与编码就是把采样后所得到的离散值经过舍入的方法变换为与输入量成正比的二进制数码。由 A/D 转换过程可以看出，它所涉及的主要技术指标包括如下几项。

（1）转换时间和转换频率。

A/D 转换器完成一次模拟量变换为数字量所需的时间即 A/D 转换时间。通常，转换频率是转换时间的倒数，它反映了采集系统的实时性能，是一个很重要的技术指标。

（2）量化误差与分辨率。

分辨率是指 A/D 转换器读输入电压微小变化响应能力的度量，习惯上以输出的二进制数码位数或 BCD 码位数表示。A/D 转换器的分辨率不采用可分辨的输入模拟电压相对值表示，这与一般测量仪表的分辨率表示方式不同。例如，A/D 转换器 AD574A 的分辨率为 12 位，即

该转换器的输出数据可以用 2^{12} 个二进制数进行量化。如果用百分数来表示分辨率，则

$$\frac{1}{2^{12}} \times 100\% = \frac{1}{4096} \times 100\% \approx 0.0244\%$$

当转换位数相同而输入电压的满量程值 U_{FS} 不同时，可分辨的最低电压也不同。例如，当分辨率为 12 位，U_{FS}=5V 时，可分辨的最低电压是 1.22mV；而当 U_{FS}=10V 时，可分辨的最低电压是 2.44mV。当输入电压的变化低于此值时，转换器不能分辨，如 9.998～10V 所转换的数字量均为 4095。

输出为 BCD 码的 A/D 转换器一般用位数表示分辨率，如双积分式 A/D 转换器 MC14433 的分辨率为 $3\frac{1}{2}$ 位。满度字位为 1999，用百分数表示分辨率为

$$\frac{1}{1999} \times 100\% = 0.05\%$$

量化误差与分辨率是统一的。量化误差是由于有限数字对模拟数值进行离散取值（量化）而引起的误差。因此，量化误差理论上为一个单位分辨率，即 $\pm\frac{1}{2}$ LSB，提高分辨率即可减少量化误差。

（3）转换精度。

A/D 转换器的转换精度反映了一个实际 A/D 转换器与一个理想 A/D 转换器在量化值上进行 A/D 转换的差值，可表示成绝对误差或相对误差，与一般测试仪表的定义相似。

对于不同的 A/D 转换器生产厂家，其产品精度指标表达式可能不完全相同。有的给出综合误差指标，有的给出分项误差指标。分项误差指标通常有非线性误差、零点误差和增益误差等。

注意：A/D 转换器的转换精度所对应的误差指标是不包括量化误差的。

ADC0809 芯片的
使用微课视频

9.1.2 ADC0809 的使用

常见的逐次逼近式 A/D 转换器有 ADC0809、AD574A 等，在此仅以最简单、廉价的 ADC0809 为例进行介绍。ADC0809 是一种有 8 路模拟输入、8 位并行数字输出的逐次逼近式 A/D 转换器。

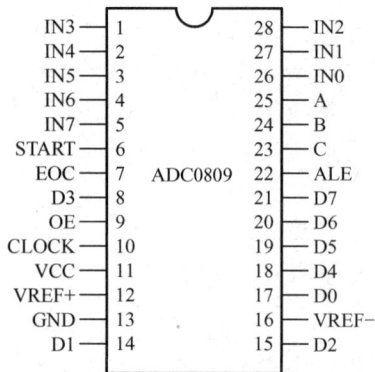

图 9-2 ADC0809 的引脚排列图

1. ADC0809 的主要技术指标和特性

（1）分辨率：8 位。

（2）转换时间：取决于芯片的时钟频率，一次 A/D 转换所需的时间。

（3）单一电源：+5V。

（4）模拟输入电压范围：单极性为 0～5V。

2. ADC0809 的引脚排列图与功能

ADC0809 的引脚排列图如图 9-2 所示。

各引脚功能如下。

（1）IN0～IN7：8 路模拟量的输入端。

（2）D0～D7：A/D 转换后的数据输入端，为三态可控输

出，可直接与计算机数据线相连。

（3）A、B、C：模拟通道地址选择端，A 为低位，C 为高位，其通道选择的地址编码如表 9-1 所示。

（4）VREF+、VREF−：基准参考电压的正、负端，决定输入模拟量的量程范围，可用单一电源供电。如果 VREF+接 5V，VREF−接地，则输入电压范围为 0～5V，此时的数字量变化范围为 0～255；如果输入电压范围为 0～2V，但希望得到的数字量变化范围还是 0～255，则可采取使 VREF+接 2V，VREF−仍然接地的方法。

（5）CLOCK：时钟信号输入端，决定 A/D 转换速率，时钟信号频率范围为 50～800kHz。

（6）ALE：地址锁存允许信号，高电平有效。当此信号有效时，A、B、C 地址信号被锁存，译码选通对应模拟通道。

（7）START：启动转换信号，正脉冲有效，通常与 $\overline{WR}$ 信号相连，控制启动 A/D 转换。

（8）EOC：转换结束信号，高电平有效。表示一次 A/D 转换已完成，可作为中断触发信号，也可用程序查询的方法检测转换是否完成。

（9）OE：输出允许信号，高电平有效，可与 $\overline{RD}$ 信号相连。当计算机发出此信号时，ADC0809 的三态门被打开，此时可通过数据线读到正确的转换结果。

表 9-1　ADC0809 模拟通道选择的地址编码

地址编码			模拟通道号	地址编码			模拟通道号
C	B	A		C	B	A	
0	0	0	IN0	1	0	0	IN4
0	0	1	IN1	1	0	1	IN5
0	1	0	IN2	1	1	0	IN6
0	1	1	IN3	1	1	1	IN7

3. ADC0809 的原理结构

ADC0809 的原理结构框图如图 9-3 所示。

由图 9-3 可知，ADC0809 内部主要包括 4 部分，各部分的主要作用如下。

（1）多路模拟量开关由于选择进入 ADC0809 的模拟通道信号，因此最多允许 8 路模拟量分时输入，公用 1 个逐次逼近式 A/D 转换器进行转换，这是一种经济的多路数据采集方法。

（2）8 路模拟量开关的切换由地址锁存与译码电路控制，A、B、C 通过 ALE 锁存，改变 A、B、C 的状态，就可以切换 8 路模拟通道，选择不同的模拟量输入。

（3）A/D 转换结果通过三态输出锁存器输出，因此，系统连接时，允许直接与单片机的数据总线相连。

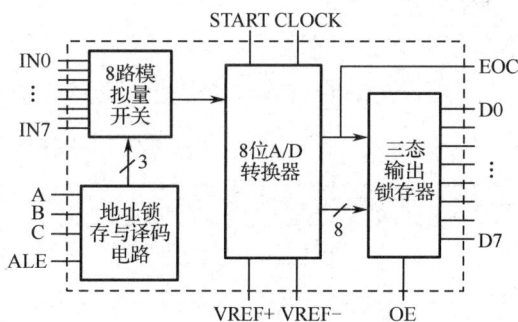

图 9-3　ADC0809 的原理结构框图

9.1.3 ADC0809 数字电压表的任务实施

【设计要求】

使用 AT89C51 和 ADC0809 设计一个数字电压表，采用数码管显示采集的电压，由于 Proteus 仿真软件不支持对 ADC0809 的仿真，因此在绘制仿真图时，采用 ADC0808 代替 ADC0809，如图 9-4 所示。

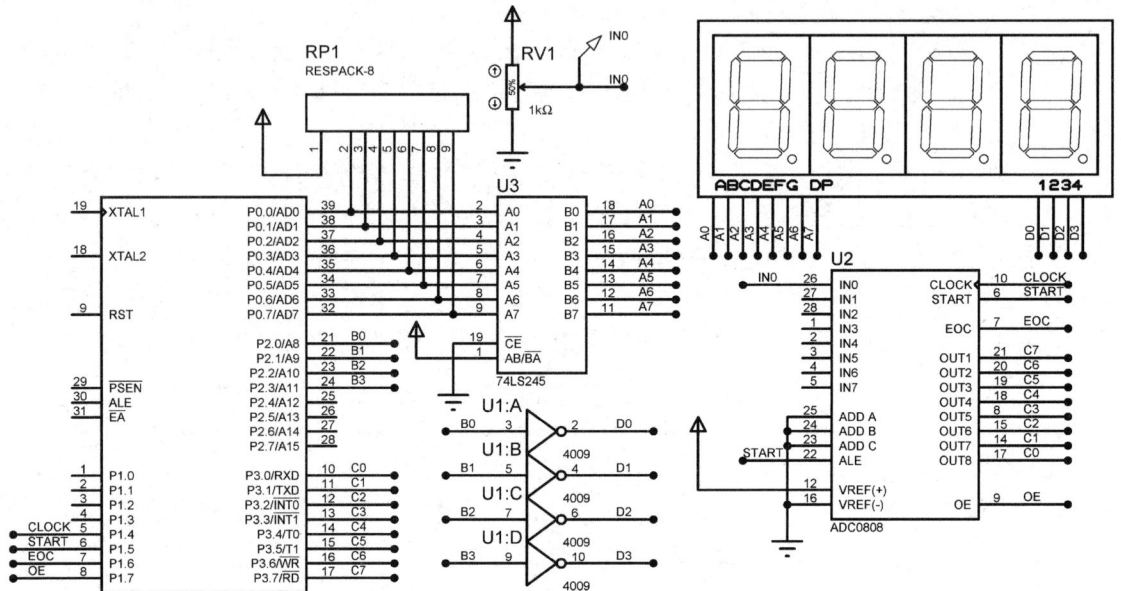

图 9-4 ADC0809 数字电压表仿真图

【任务分析】

根据设计要求，采用四位一体共阳极数码管作为显示器件，单片机产生的 50kHz 方波作为 ADC0808 的时钟信号，在仿真时，通过调节 RV1 电位器来实现不同电压的采集。

【实施步骤】

1. 添加元器件

打开 Proteus 仿真软件，按照表 9-2 添加元器件。

表 9-2 ADC0809 数字电压表的元器件清单

元器件名称	所属类	所属子类
AT89C51	Microprocessor ICs	8051 Family
RES	Resistors	Generic
7SEG-MPX4-CA	Optoelectronics	7-Segment Displays
RESPACK-8	Resistors	Resistor Packs
74LS245	TTL 74LS series	Transceivers
4009	CMOS 4000 series	Buffers & Drivers
ADC0808	Data Converters	A/D Converters
POT-HG	Resistors	Variable

2. 绘制仿真图

元器件全部添加后，在 Proteus ISIS 的原理图编辑窗口中按图 9-4 绘制 ADC0809 数字电压表仿真图。

3. 编写程序

ADC0808 的时钟信号采用 50kHz 的方波，可以通过定时器 T0 中断产生此方波信号。当启动 ADC0808 的 A/D 转换后，可以通过检测 EOC 引脚的值来判断转换是否完成。

在 Keil μVision5 中编写程序，实现 ADC0809 数字电压表的效果，参考程序如下：

```
/*********************************************************
程序名称：program9-1.c
程序功能：ADC0809 数字电压表程序
*********************************************************/
#include<reg52.h>                        //加载头文件
#include<intrins.h>                      //加载头文件
/*********************************************************
数据类型定义
*********************************************************/
#define uchar unsigned char              //定义无符号字符型
#define uint unsigned int                //定义无符号整型
/*********************************************************
单片机引脚定义
*********************************************************/
sbit CLOCK=P1^4;
sbit START=P1^5;
sbit EOC=P1^6;
sbit OE=P1^7;
/*********************************************************
全局变量定义
*********************************************************/
uchar AD_DATA=0x00;
/*********************************************************
数组定义：共阳极数码管的编码表
*********************************************************/
uchar code LED_7SEG_CA[]=
{    //0    1    2    3    4    5    6    7    8    9
    0xc0, 0xf9, 0xa4, 0xb0, 0x99, 0x92, 0x82, 0xf8, 0x80, 0x90
};
/*********************************************************
数组定义：显示单元，4 字节
*********************************************************/
uchar DISPLAY_DATA[4]={0};
/*********************************************************
延时 t ms 函数
*********************************************************/
```

ADC0809 数字电压表参考程序

```
void delayms(uchar t)
{
    uchar i;
    while(t--)
        for(i=96;i>0;i--);
}
```
/***
显示函数
***/
```
void display()
{
    uchar i=0xfe, j;
    for(j=0;j<4;j++)
    {
        P2=i;
        P0=DSP_7SEG_CODE[DISPLAY_DATA[j]];
        if(j==0)
            P0=P0&0x7f;                          //点亮小数点
        delayms(1);
        i=_crol_(i,1);
    }
}
```
/***
ADC0808 驱动函数
***/
```
void adc_init()
{
    START=0;
    START=1;
    START=0;
    while(EOC==0){}                              //判断转换是否完成
    OE=1;
    AD_DATA=P3;
}
```
/***
函数名：ZD_T0 ()
功能描述：定时器 T0 中断产生 50kHz 的方波
***/
```
void ZD_T0()interrupt 1
{
    CLOCK=~CLOCK;
}
```
/***
函数名：void main()
功能描述：主函数

```
************************************************************************/
void main()
{
    uint i;
    uchar j;
    TMOD=0x02;
    TH0=246;
    TL0=246;
    EA=1;
    ET0=1;
    TR0=1;
    while(1)
    {
        adc_init();
        i=AD_DATA*196;
        DISPLAY_DATA[0]=i/10000;
        DISPLAY_DATA[1]=i%10000/1000;
        DISPLAY_DATA[2]=i%1000/100;
        DISPLAY_DATA[3]=i%100/10;
        for(j=10;j>0;j--)
            display();
    }
}
```

4. 系统仿真

当 Keil C51 编译成功后，会自动产生 HEX 文件，接着打开之前绘制的 Proteus 仿真图，双击 AT89C2051，弹出"Edit Component"对话框，单击"Program File"中的文件夹按钮，在弹出的"Select File Name"对话框中，选择之前编译生成的 HEX 文件，单击"打开"按钮，返回"Edit Component"对话框，单击"OK"按钮，即可装入 HEX 文件。

接着单击 Proteus ISIS 编辑界面左下角的运行按钮 ▶，即可观察是否能够实现 ADC0809 数字电压表的显示效果，如图 9-5 所示。

ADC0809 数字
电压表仿真
效果视频

图 9-5　ADC0809 数字电压表仿真效果图

任务 9.2　ADC0831 数字电压表的仿真设计

学习目标

ADC0831 数字电压表导学材料

【知识目标】

（1）了解并掌握 ADC0831 的基本原理和使用方法。

（2）了解并掌握 LCD1602 的使用方法。

【技能目标】

（1）了解并掌握单片机仿真软件 Proteus 的使用方法。

（2）了解并掌握单片机编译软件 Keil C51 的使用方法。

（3）通过 ADC0831 数字电压表的仿真设计初步了解并掌握单片机项目的开发步骤。

【思政目标】

在 ADC0831 数字电压表的仿真设计中，展现我国在液晶显示技术领域的进步和成就，激发学生的爱国情怀和民族自信心。

ADC0831 的使用

9.2.1　ADC0831 的使用

1. ADC0831 的主要技术指标和特性

ADC0831 是美国国家半导体公司推出的一种逐次逼近式串行 A/D 转换器，其具有以下特点。

① ADC0831 是一款 8 位 A/D 转换器。

② ADC0831 可通过三线串行总线与单片机连接。

③ ADC0831 是单通道的 A/D 转换器。

④ ADC0831 可以单端输入，也可以采用差分输入。

⑤ ADC0831 最大功耗为 0.8W。

⑥ ADC0831 电源电压最大值为 6.3V，最小值为 4.5V。

⑦ ADC0831 最高工作温度为+70℃。

⑧ ADC0831 输入电压为 5V，参考电压为 5V。

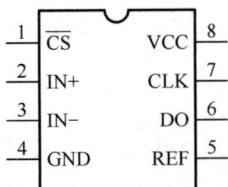

图 9-6　ADC0831 的引脚排列图

ADC0831 的引脚排列图如图 9-6 所示，其中 1 脚为片选端（$\overline{\text{CS}}$），2 脚为正输入信号端（IN+），3 脚为负输入信号端（IN−），4 脚为地（GND），5 脚为参考电压输入端（REF），6 脚为串行数据输出端（DO），7 脚为时钟信号输入端（CLK），8 脚为电源端（VCC）。

从图 9-7 中可以看出，当 $\overline{\text{CS}}$ 为低电平后，ADC0831 被选中，此时 CLK 输入 2 个时钟信号，ADC0831 将前一次转换的结果中的最高有效位（MSB）通过 DO 输出，接着要求 CLK 继续输入 8 个时钟信号，单片机就可以通过 ADC0831 的 DO 读取 A/D 转换数据。

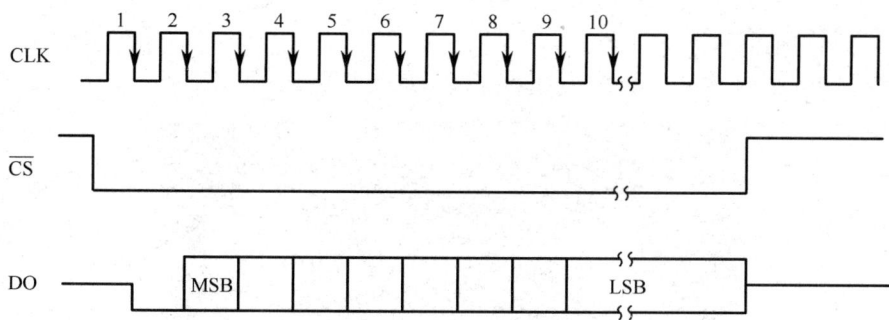

图 9-7　ADC0831 时序图

2. ADC0831 的 A/D 转换函数设计

ADC0831 的 A/D 转换函数流程图如图 9-8 所示，我们在编写程序时，必须严格按照图 9-7 所示的 ADC0831 时序图进行编程，否则 ADC0831 可能不会正常工作。

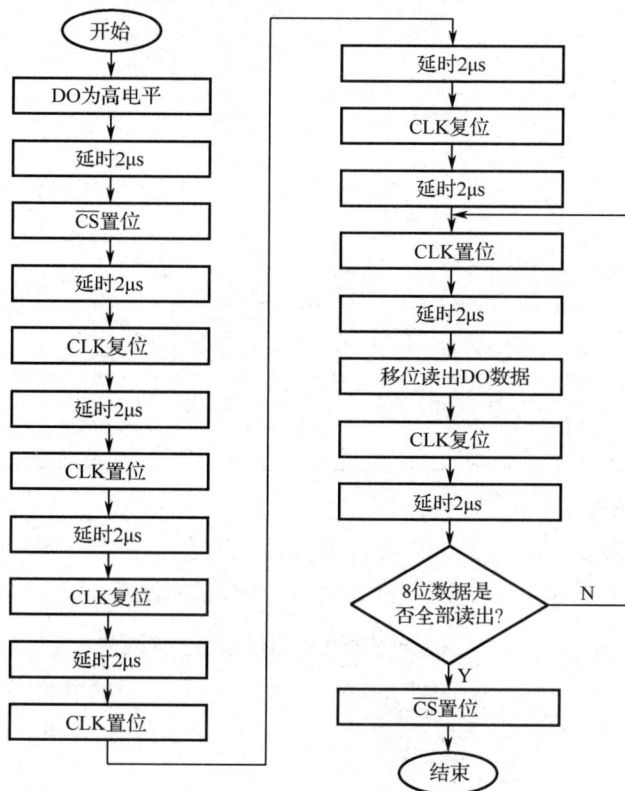

图 9-8　ADC0831 的 A/D 转换函数流程图

9.2.2　LCD1602 的使用

LCD1602 液晶屏
的使用微课视频

LCD（液晶屏）是一种用液晶材料制成的功耗极低的显示器件，具有电压低、功耗极低、体积小、寿命长、被动显示、没有电磁辐射等特点，是便携式和手持仪器仪表的首选显示器。LCD 可分为笔端型、字符型、点阵图形型等，常见的字符型 LCD 主要有 LCD1602 等。

1. LCD1602 概述

LCD1602 的外形如图 9-9 所示，引脚功能如表 9-3 所示。

图 9-9　LCD1602 的外形

表 9-3　LCD1602 的引脚功能

引脚号	引脚名称	功能
1	VSS	接地
2	VDD	电源正极
3	VO	液晶显示对比度调整端，可连接电位器
4	RS	数据/命令选择位，RS=0：命令；RS=1：数据
5	R/$\overline{W}$	读/写选择位，R/$\overline{W}$=0：写操作；R/$\overline{W}$=1：读操作
6	E	数据读写操作控制位，其向 LCD 发送一个脉冲，LCD 与单片机之间将进行一次数据交换
7~14	DB0~DB7	8 位数据线，可以用 8 位连接，也可以只用高 4 位连接，节约单片机资源
15	A	背光控制正电源
16	K	背光控制地

2. LCD1602 基本操作

单片机对 LCD 编程控制主要有 4 种基本操作：写命令、读状态、写数据和读数据，由 LCD1602 的 3 个控制引脚 RS、R/$\overline{W}$、E 的不同组合状态确定，如表 9-4 所示。

表 9-4　LCD1602 的 3 个控制引脚状态对应的基本操作

控制引脚			基本操作
RS	R/$\overline{W}$	E	
0	0	⊓	写命令操作：用于初始化、清屏、光标定位等
0	1	⊓	读状态操作：读忙标志，当忙标志为 1 时，表明 LCD 正在进行内部操作，这时不能进行其他操作；当忙标志为 0 时，表明 LCD 内部操作已经结束，可以进行其他操作，一般采用查询方式
1	0	⊓	写数据操作：写入要显示的内容
1	1	⊓	读数据操作：将显示存储区中的数据反读出来，一般比较少用

图 9-10 给出了 LCD 读操作和写操作时序图。在读操作时，E 的高电平有效。所以在软件

设置顺序上，先设置 RS 和 R/$\overline{\text{W}}$，再设置 E 为高电平，这时从数据接口读取数据，然后设置 E 为低电平，最后复位 RS 和 R/$\overline{\text{W}}$。在写操作时，E 的下降沿有效。所以在软件设置顺序上，先设置 RS 和 R/$\overline{\text{W}}$，再设置数据，然后产生 E 的脉冲，最后复位 RS 和 R/$\overline{\text{W}}$。

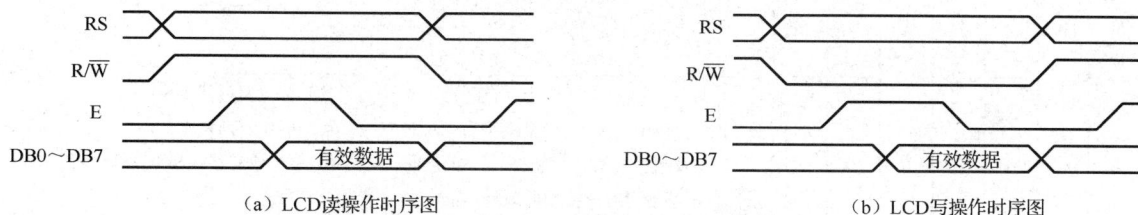

（a）LCD读操作时序图　　　　　　　　　　　　（b）LCD写操作时序图

图 9-10　LCD1602 操作时序图

3. LCD1602 控制指令

LCD1602 共有 11 条控制指令，其常用的指令格式与功能如下。

（1）数据指针设置。

控制器内部设有一个数据指针，用户可以通过它访问控制器内部的全部 80 字节的 RAM，数据指针设置如表 9-5 所示。

表 9-5　数据指针设置

指令码	功能
0x80+0x 地址码（0～27H，40H～67H）	设置数据指针

（2）清屏与回车指令设置。

清屏与回车指令设置如表 9-6 所示。

表 9-6　清屏与回车指令设置

指令码	功能
0x01	显示清屏，1：数据指针清零；2：所有显示清零
0x02	显示回车，1：数据指针清零

（3）显示模式设置。

显示模式设置如表 9-7 所示。

表 9-7　显示模式设置

指令码								功能
0	0	1	1	1	0	0	0	设置 16×2 显示，5×7 点阵，8 位数据接口
0x38								

（4）显示开/关及光标设置。

显示开/关及光标设置如表 9-8 所示。

表 9-8 显示开/关及光标设置

指令码								功能
0	0	0	0	1	D	C	B	B=1：光标闪烁；B=0：光标不闪烁。 C=1：显示光标；C=0：不显示光标。 D=1：开显示；D=0：关显示
0	0	0	0	0	1	N	S	N=1：当读/写一个字符后地址指针加 1 且光标位置后移 1 位； N=0：当读/写一个字符后地址指针减 1 且光标位置前移 1 位。 S=1：当写一个字符后，整屏显示左移（N=1）或右移（N=0），以得到光标 不移动而屏幕移动的效果； S=0：当写一个字符后，整屏显示不移动
0	0	0	1	0	0	0	0	光标左移
0	0	0	1	0	1	0	0	光标右移
0	0	0	1	1	0	0	0	整屏左移，同时光标跟随移动
0	0	0	1	1	1	0	0	整屏右移，同时光标跟随移动

9.2.3 ADC0831 数字电压表的任务实施

【设计要求】

通过 AT89C51 和 ADC0831 设计一个数字电压表，采用 LCD1602 显示采集的电压，仿真图如图 9-11 所示。

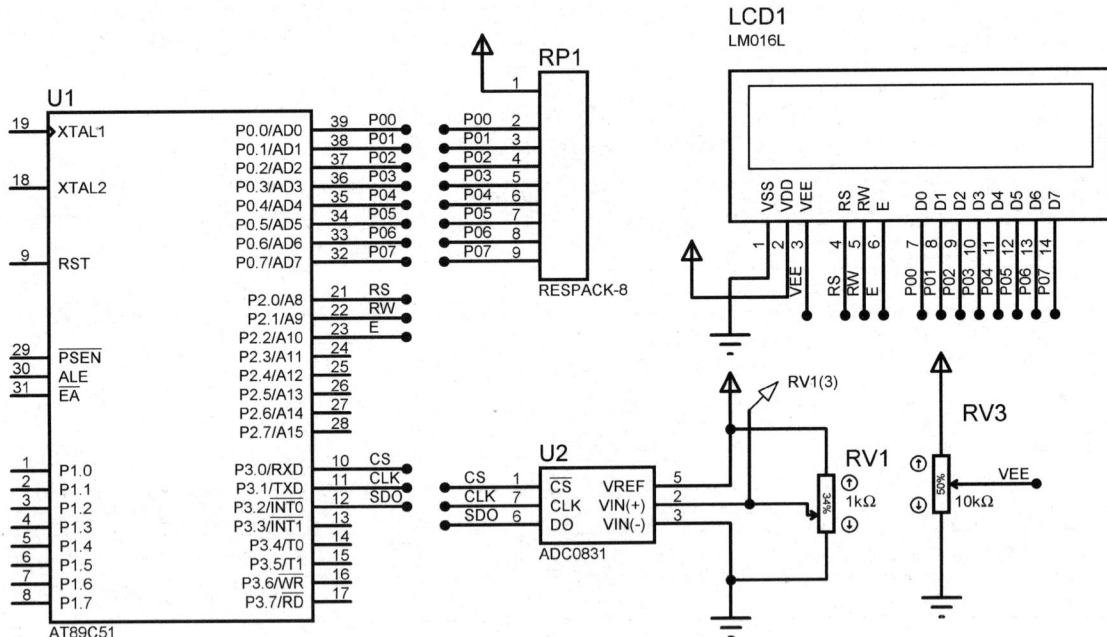

图 9-11 ADC0831 数字电压表仿真图

【任务分析】

在本任务中，采用 LCD1602 进行显示。ADC0831 的 VREF 端接+5V，$\overline{CS}$、CLK、DO

端分别接单片机的 P3.0、P3.1、P3.2 口。因为 ADC0831 为 8 位 A/D 转换器，参考电压为 5V，分辨的最低电压是 19.6mV，所以在 ADC0831 转换电压程序中，将常数 196 拆分为 49×4，即首先连续采集 49 次 A/D 转换器的值，累加后乘以 4，实现电压转换。

【实施步骤】

1. 添加元器件

打开 Proteus 仿真软件，按照表 9-9 添加元器件。注意：用 Proteus 仿真软件绘制单片机仿真图时，可以省略振荡电路和复位电路。

表 9-9　ADC0831 数字电压表的元器件清单

元器件名称	所属类	所属子类
AT89C51	Microprocessor ICs	8051 Family
RES	Resistors	Generic
LM016L	Optoelectronics	Alphanumeric LCDs
RESPACK-8	Resistors	Resistor Packs
ADC0831	Data Converters	A/D Converters
POT-HG	Resistors	Variable

2. 绘制仿真图

元器件全部添加后，在 Proteus ISIS 的原理图编辑窗口中按图 9-11 绘制 ADC0831 数字电压表仿真图。

ADC0831 数字电压表参考程序

3. 编写程序

在 Keil μVision5 中编写程序，实现 ADC0831 数字电压表的效果，参考程序如下：

```
/*****************************************************************
程序名称：program9-2.c
程序功能：ADC0831 数字电压表程序
*****************************************************************/
#include<reg52.h>                    //加载头文件
#include<intrins.h>                  //加载头文件
/*****************************************************************
数据类型定义
*****************************************************************/
#define uchar unsigned char          //定义无符号字符型
#define uint unsigned int            //定义无符号整型
/*****************************************************************
单片机引脚定义
*****************************************************************/
sbit LCD1602_RS=P2^0;                //LCD1602 数据/命令选择位
sbit LCD1602_RW=P2^1;                //LCD1602 读/写选择位
sbit LCD1602_E=P2^2;                 //LCD1602 数据读写操作控制位
sbit LCD1602_IO=P0^7;                //LCD1602 最高位数据位
sbit CS=P3^0;                        //ADC0831 片选端
```

```
sbit CLK=P3^1;                              //ADC0831 时钟信号输入端
sbit SDO=P3^2;                              //ADC0831 串行数据输出端
/**********************************************************************
全局变量定义
**********************************************************************/
uchar LCD1602_DATA;                         //LCD1602 待写入数据存储单元
uchar ADC_DATA=0;                           //ADC0831 数据存储单元
uchar DISPLAY_DATA[4]={0};                  //显示存储单元
/**********************************************************************
函数名：void lcd1602_delay()
功能描述：LCD1602 延时函数
**********************************************************************/
void lcd1602_delay()
{
Loop: P0=0xff;
     LCD1602_RS=0;
     LCD1602_RW=1;
     LCD1602_E=0;
     _nop_();
     LCD1602_E=1;
     if(LCD1602_IO==1)
          goto loop;
}
/**********************************************************************
函数名：void lcd1602_enable(uchar i)
功能描述：LCD1602 写命令函数
**********************************************************************/
void lcd1602_enable(uchar i)
{
     P0=i;                                  //通过 P0 口输出数据
     LCD1602_RS=0;                          //RS 复位
     LCD1602_RW=0;                          //RW 复位
     LCD1602_E=0;                           //E 复位
     lcd1602_delay();
     LCD1602_E=1;                           //E 置位
}
/**********************************************************************
函数名：void lcd1602_init()
功能描述：LCD1602 初始化函数
**********************************************************************/
void lcd1602_init()
{
     lcd1602_enable(0x01);                  //清屏并光标复位
     lcd1602_enable(0x38);                  //设置显示模式，采用 8 位 2 行 5×7 点阵
     lcd1602_enable(0x0c);                  //显示器开、光标开、光标允许闪烁
```

```
    lcd1602_enable(0x06);                      //文字不动，光标自动右移
}
/*****************************************************************
函数名：void lcd1602_write()
功能描述：LCD1602 写数据函数
*****************************************************************/
void lcd1602_write()
{
    LCD1602_RS=1;                              //RS 置位
    LCD1602_RW=0;                              //RW 复位，准备写入数据
    LCD1602_E=1;
    LCD1602_E=0;                               //E 清零，执行显示命令
    lcd1602_delay();
    LCD1602_E=1;                               //E 置位，显示完成
}
/*****************************************************************
函数名：void display()
功能描述：显示函数
*****************************************************************/
void display()
{
    lcd1602_enable(0x80);                      //写入显示起始地址（第 1 行第 0 个位置）
    P0='U';                                    //显示字符"U"
    lcd1602_write();                           //LCD1602 写数据函数
    P0=':';                                    //显示字符":"
    lcd1602_write();                           //LCD1602 写数据函数
    LCD1602_DATA=DISPLAY_DATA[0];              //显示电压的整数部分
    P0=LCD1602_DATA+0x30;                      //转 ASCII 码送显
    lcd1602_write();                           //LCD1602 写数据函数
    P0='.';                                    //显示字符"."
    lcd1602_write();                           //LCD1602 写数据函数
    LCD1602_DATA=DISPLAY_DATA[1];              //显示电压的小数后 1 位
    P0=LCD1602_DATA+0x30;                      //转 ASCII 码送显
    lcd1602_write();                           //LCD1602 写数据函数
    LCD1602_DATA=DISPLAY_DATA[2];              //显示电压的小数后 1 位
    P0=LCD1602_DATA+0x30;                      //转 ASCII 码送显
    lcd1602_write();                           //LCD1602 写数据函数
    LCD1602_DATA=DISPLAY_DATA[3];              //显示电压的小数后 1 位
    P0=LCD1602_DATA+0x30;                      //转 ASCII 码送显
    lcd1602_write();                           //LCD1602 写数据函数
    P0='V';                                    //显示字符"V"
    lcd1602_write();                           //LCD1602 写数据函数
}
/*****************************************************************
函数名：void adc_0831()
```

功能描述：adc_0831 驱动函数
**/

```c
void adc_0831()
{
    uchar i;
    SDO=1;
    _nop_();
    _nop_();
    CS=0;
    _nop_();
    _nop_();
    CLK=0;
    _nop_();
    _nop_();
    CLK=1;
    _nop_();
    _nop_();
    CLK=0;
    _nop_();
    _nop_();
    CLK=1;
    _nop_();
    _nop_();
    CLK=0;
    _nop_();
    _nop_();
    for(i=0;i<8;i++)
    {
        CLK=1;
        _nop_();
        _nop_();
        ADC_DATA<<=1;
        if(SDO==1)
            ADC_DATA++;
        CLK=0;
        _nop_();
        _nop_();
    }
    CS=1;
    _nop_();
    _nop_();
}
```
/**
函数名：void main()
功能描述：主函数

```
*************************************************************************/
void main()
{
    uchar i;
    uint j;
    lcd1602_init();                          //调用 LCD1602 初始化函数
    while(1)
    {
        j=0;
        for(i=49;i>0;i--)                    //连续采集 49 次并累加
        {
            adc_0831();
            j=j+ADC_DATA;
        }
        j=j*4;                               //49 次采集值累加并乘以 4，实现电压转换
        DISPLAY_DATA[0]=j/10000;
        DISPLAY_DATA[1]=j%10000/1000;
        DISPLAY_DATA[2]=j%1000/100;
        DISPLAY_DATA[3]=j%100/10;
        display();
        LED=~LED;
    }
}
```

4. 系统仿真

当 Keil C51 编译成功后，会自动产生 HEX 文件，接着打开之前绘制的 Proteus 仿真图，双击 AT89C51，弹出"Edit Component"对话框，单击"Program File"中的文件夹按钮，在弹出的"Select File Name"对话框中，选择之前编译生成的 HEX 文件，单击"打开"按钮，返回"Edit Component"对话框，单击"OK"按钮，即可装入 HEX 文件。

ADC0831 数字电压表仿真效果视频

接着单击 Proteus ISIS 编辑界面左下角的运行按钮 ▶ ，即可观察是否能够实现 ADC0831 数字电压表的功能，如图 9-12 所示。

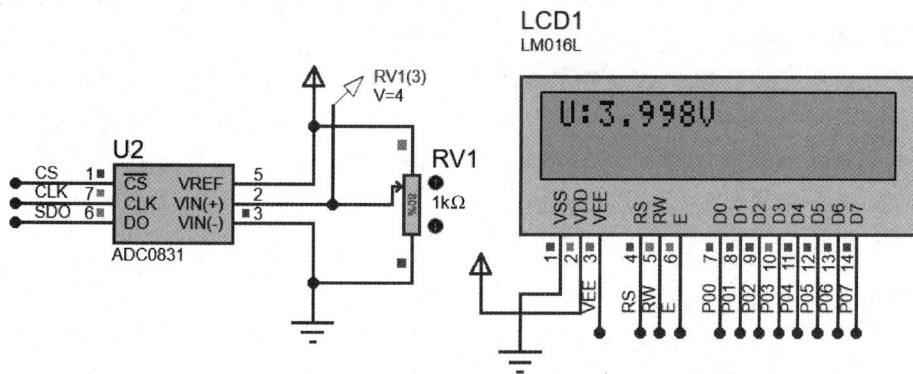

图 9-12　ADC0831 数字电压表仿真效果图

素养小课堂

单片机程序员的三大工匠精神

单片机程序员的三大工匠精神——细节决定成败、敬畏程序、善于总结，对单片机学习者来说是极其宝贵的指导原则。这些精神不仅能够帮助他们在技术道路上稳步前行，还能激发他们不断思考、创新，并追求极致的程序设计。

1. 细节决定成败

深入理解硬件特性：单片机程序设计离不开对硬件的深入理解。学生应详细阅读数据手册，了解每个引脚的功能、寄存器的配置及外设的工作原理，确保在编程时能精确控制硬件行为。

代码优化与调试：注重代码中的每个细节，包括变量命名、注释、代码结构等。通过代码审查、单元测试等手段，确保程序逻辑正确，无内存泄漏、死循环等潜在问题。同时，利用调试工具逐步跟踪程序执行过程，解决细微的 Bug。

边界条件测试：在设计程序时，特别关注输入和输出的边界条件，确保程序在极端情况下仍能稳定运行。

2. 敬畏程序

尊重编程规范：遵循行业公认的编程规范和最佳实践，如代码风格指南、命名规范等，使代码易于阅读和维护。

严谨的逻辑思考：在编写程序前，先通过流程图、伪代码等方式规划好程序的整体结构和逻辑流程，确保每步都经过深思熟虑。

持续学习与反思：保持对新技术、新方法的敏感度，不断学习并应用到实践中。同时，对过去的项目进行总结和反思，找出不足之处并改进。

3. 善于总结

项目文档化：为每个项目建立详细的文档，包括项目背景、需求分析、设计方案、实现过程、测试结果等。这有助于团队成员之间的沟通和协作，也为未来的项目提供参考。

经验分享与交流：积极参与技术社区、论坛等平台的讨论，分享自己的经验和心得。同时，倾听他人的意见和建议，从中汲取营养。

知识沉淀与复用：将常用的函数、模块等封装成库或框架，以便在未来的项目中复用。这不仅可以提高工作效率，还能减少重复劳动和错误。

在单片机程序设计领域，不仅要掌握扎实的理论基础和编程技能，还要具备创新思维和解决问题的能力。通过不断思考、创新、反复打磨和追求极致，单片机学习者才能够找到最佳的单片机程序设计思路，并在实践中不断验证和完善自己的想法，最终成为这个领域的佼佼者，为科技进步和社会发展贡献自己的力量。

课后任务

1. 在任务 9.1 的基础上添加功能，采用 ADC0808 和单片机设计一个能够同时采集两路电压的数字电压表，要求每隔 3s 切换一次 A/D 转换模拟通道，采集不同的电压，如图 9-13

所示。显示效果参见二维码。

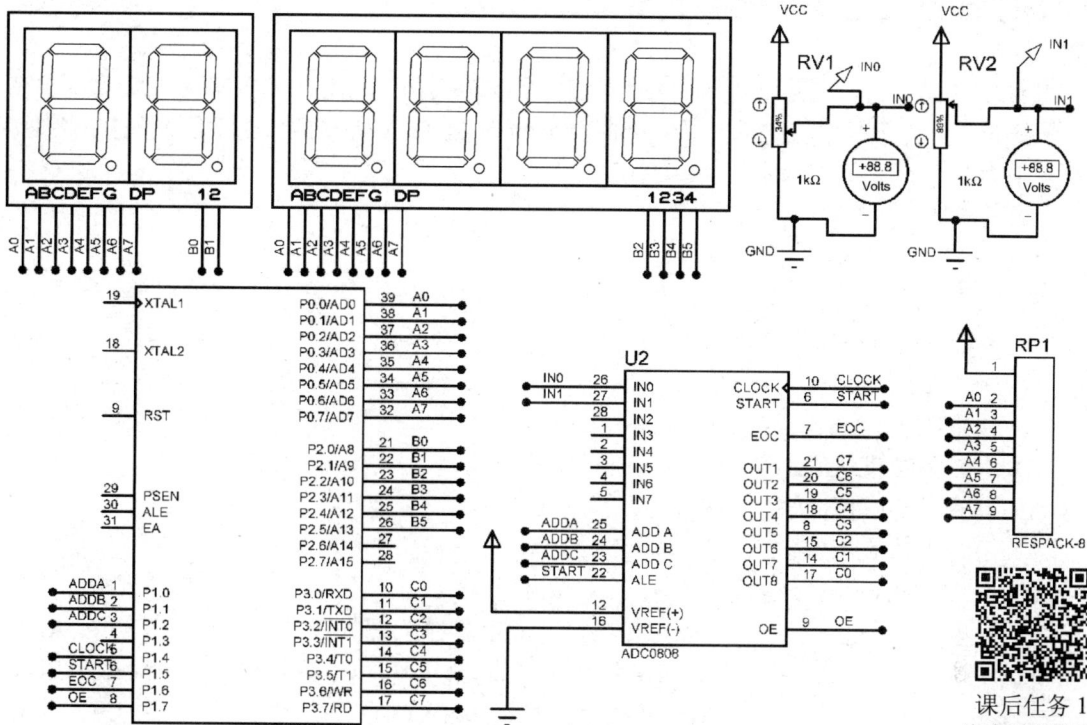

图 9-13　课后任务 1 电路图

课后任务 1
仿真效果视频

课后任务 2
仿真效果视频

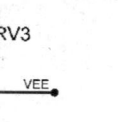

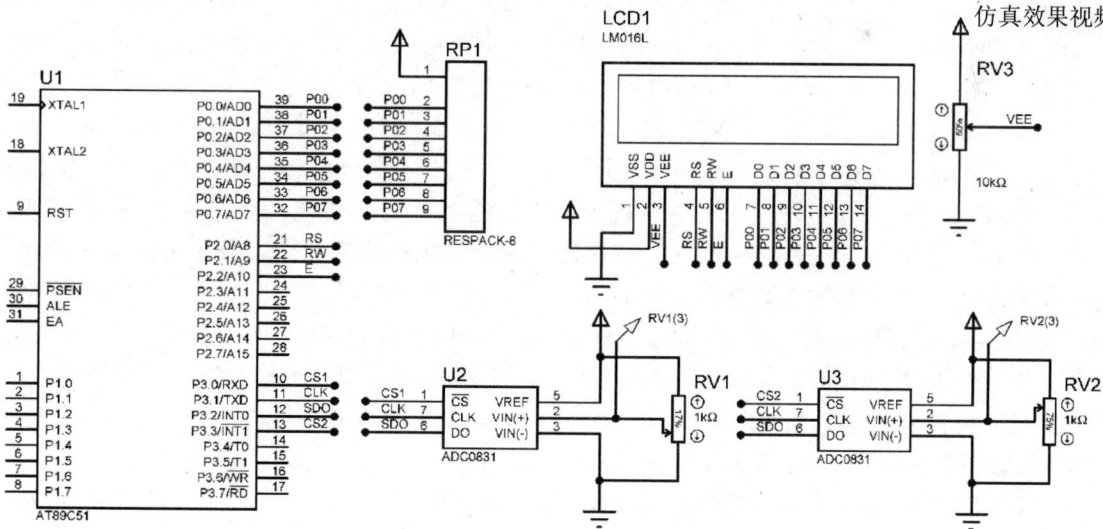

2．在任务 9.2 的基础上添加功能，采用 2 块 ADC0831 和单片机设计一个能够同时采集两路电压的数字电压表，如图 9-14 所示。显示效果参见二维码。

图 9-14　课后任务 2 电路图

知识拓展 串行 A/D 转换器 TLC2543

TLC2543 是 TI 公司推出的一种 12 位串行 A/D 转换器，使用开关电容逐次逼近技术完成 A/D 转换过程。由于是串行输入结构，因此能够节省 51 单片机的 I/O 接口资源，且价格适中。其特点主要如下。

（1）分辨率为 12 位。

（2）在工作温度范围内转换时间为 10μs。

（3）具有 11 路模拟输入通道。

（4）3 路内置自测试方式。

（5）采样率为 66kHz。

（6）线性误差+1LSB（max）。

（7）有转换结束（EOC）输出。

（8）具有单、双极性输出。

（9）可编程的 MSB 或 LSB 前导。

（10）可编程的输出数据长度。

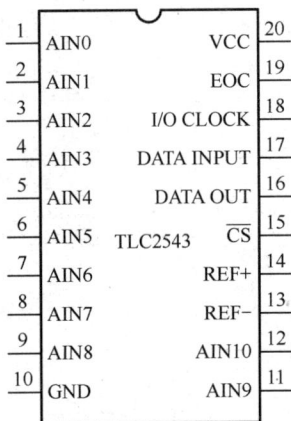

图 9-15 TLC2543 的引脚排列图

TLC2543 的引脚排列图如图 9-15 所示，其中 AIN0～AIN10 为 11 路模拟输入端，$\overline{CS}$ 为片选端，DATA INPUT 为串行数据输入端，DATA OUT 为 A/D 转换结果的三态串行输出端，EOC 为转换结束端，I/O CLOCK 为 I/O 时钟端，REF+为正基准电压端，REF−为负基准电压端，VCC 为电源端，GND 为地。

TLC2543 的控制字为 8 位数据，从 DATA INPUT 端串行输入，它规定了 TLC2543 要转换的模拟量通道号、转换后的输出数据长度及输出数据格式，其中高 4 位（D7～D4）决定了通道号，若高 4 位数据为 1000，则通道号为 8，由于 TLC2543 共有 11 个通道，所以当高 4 位数据为 1011～1101 时，就不再表示通道号，而是作用 TLC2543 的自检、分别测试（$V_{REF+}+V_{REF-}$）/2、V_{REF+}、V_{REF-}的值，当高 4 位为 1110 时，

TLC2543 进入休眠状态。控制字的低 4 位决定了输出数据长度及格式，其中 D3、D2 决定了输出数据长度，01 表示输出数据长度为 8 位，11 表示输出数据长度为 16 位，其他表示为 12 位，D1 决定了输出数据是先输出高位，还是先输出低位，若 D1=0，则表示先输出高位，D0 决定了输出数据是单极性（二进制）还是双极性（2 的补码），若 D0=0，则表示单极性。

TLC2543 的串行总线为 SPI 总线，51 单片机没有 SPI 或相同的接口能力，需要采用软件来模拟 SPI 的时序操作，TLC2543 时序图如图 9-16 所示。

当 TLC2543 开始上电后，$\overline{CS}$ 端必须从高到低，才能开始一次工作周期，此时 EOC 端为高电平，输入数据寄存器被置 0，输出数据寄存器的内容是随机的。开始时，$\overline{CS}$ 端为高电平，I/O CLOCK、DATA INPUT 端被禁止，DATA OUT 端呈高阻态，EOC 端为高电平，然后使 $\overline{CS}$ 端为低电平，I/O CLOCK、DATA INPUT 端使能，DATA OUT 端脱离高阻态，12 个时钟信

号从 I/O CLOCK 端依次加入，随着时钟信号的加入，控制字从 DATA INPUT 端开始一位一位地在时钟信号的上升沿被送入 TLC2543（高位先送入），同时上一个工作周期转换的 A /D 数据（输出数据寄存器中的数据）从 DATA OUT 端开始一位一位地移出。TLC2543 收到第 4 个时钟信号后，通道号也已收到，此时 TLC2543 开始对选定通道的模拟量进行采样，并保持到第 12 个时钟信号的下降沿。在第 12 个时钟信号的下降沿来临时，EOC 端变为低电平，开始对本次采样的模拟量进行 A/D 转换，转换时间约需要 $10\mu s$，转换完成后 EOC 端变为高电平，转换的数据在输出数据寄存器中，待下一个工作周期输出。此后，TLC2543 可以进行新的工作周期。

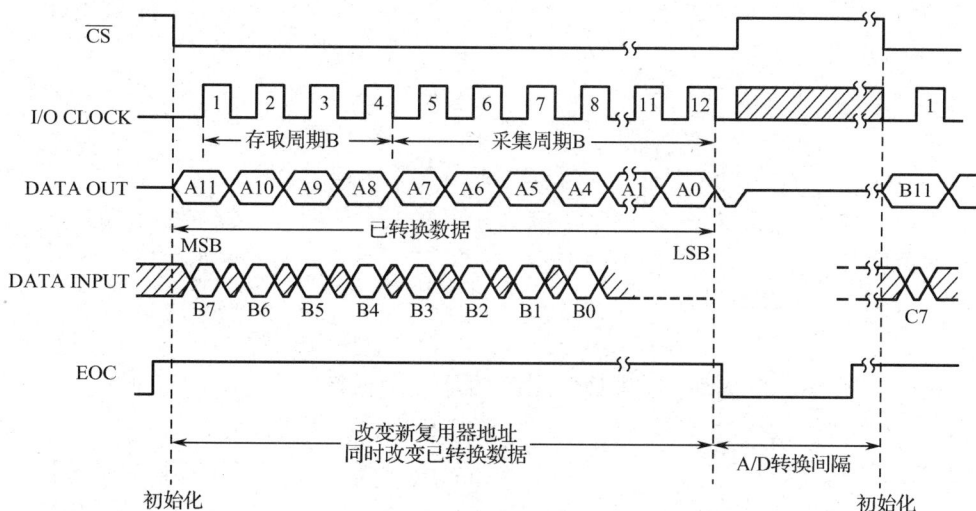

图 9-16　TLC2543 时序图

TLC2543 的 A/D 转换函数流程图如图 9-17 所示，TLC2543 在每次 I/O 周期读取的数据都是上次转移的结果，当前的转换结果在下一个 I/O 周期中被串行移出。由于内部调整，因此第一次读数读取的转换结果可能不准确，应丢弃。

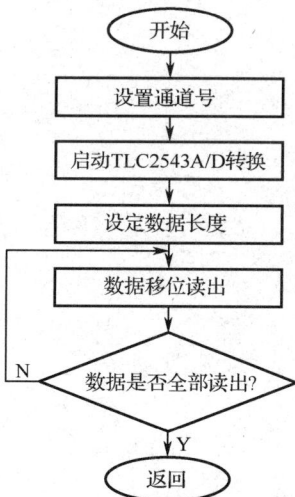

图 9-17　TLC2543 的 A/D 转换函数流程图

习题

一、单选题

1. ADC0809 是（　　）芯片。

A. 8 位 D/A 转换器 　　　　　　　　　　 B. 16 位 D/A 转换器

C. 8 位 A/D 转换器 　　　　　　　　　　 D. 16 位 A/D 转换器

2. ADC0809 的转换启动信号为（　　）。

A. EOC 　　　　　 B. XFER 　　　　　 C. ILE 　　　　　 D. START

3. A/D 转换可以实现（　　）功能。

A. 模拟信号转换为数字信号 　　　　　　 B. 数字信号转换为模拟信号

C. 编码 　　　　　　　　　　　　　　　 D. 数据选择

4. LCD1602 一行可以显示（　　）个字符。

A. 8 　　　　　　　 B. 16 　　　　　　 C. 24 　　　　　　 D. 32

二、多选题

1. LCD1602 的优点主要有（　　）。

A. 功耗低 　　　　　 B. 能显示中文 　　　 C. 能显示英文 　　　 D. 亮度高

2. 关于 LCD1602 的引脚，以下说法正确的是（　　）。

A. D0～D7 为数据端 　　　　　　　　　　 B. A 为背光源负极

C. E 为使能信号端 　　　　　　　　　　　 D. VO 为液晶显示对比度调整端

三、判断题

1. 在 A/D 转换器中，逐次逼近式在精度上不及双积分式，但双积分式在速度上较低。

（　　）

2. A/D 转换的精度不仅取决于量化位数，还取决于参考电压。　　　　　　（　　）

3. 在 A/D 转换时，转换频率越高越好。　　　　　　　　　　　　　　　（　　）

项目十　信号发生器的仿真设计

任务 10.1　DAC0832 正弦波信号发生器的仿真设计

学习目标

【知识目标】

（1）了解并掌握 D/A 转换的原理。

（2）了解并掌握 DAC0832 的使用方法。

【技能目标】

（1）了解并掌握单片机仿真软件 Proteus 的使用方法。

（2）了解并掌握单片机编译软件 Keil C51 的使用方法。

（3）了解并掌握单片机程序下载的方法。

（4）了解并掌握单片机最小系统的组成。

（5）通过 DAC0832 正弦波信号发生器的仿真设计初步了解并掌握单片机项目的开发步骤。

DAC0832 正弦波信号发生器导学材料

【思政目标】

通过 DAC0832 正弦波信号发生器的仿真设计，引导学生思考如何优化正弦波信号发生器的性能，提高其稳定性和精度。

什么是 D/A 转换微课视频

10.1.1　D/A 转换

在单片机控制系统中，很多控制对象用的是模拟量，如对电机、机械手、记录仪等设备的控制等，所以必须将单片机输出的数字量转换为模拟电压或电流，送到执行机构以达到某种控制过程，所有这些都离不开 D/A 转换接口。此外，D/A 转换还可以产生各种波形，所以 D/A 转换接口是数字化测控系统及智能仪器中的必要组成部分。

1. D/A 转换的原理

D/A 转换是将数字量转换为与此数值成正比的模拟量。一个二进制数是由各位代码组合起来的，每位代码在二进制数中的位置代表一定的权。为了将数字量转换为模拟量，应将每位代码按权大小转换为相应的模拟输出分量，根据叠加原理将各代码对应的模拟输出分量相加，其总和就是与数字量成正比的模拟量，至此 D/A 转换完成。

为实现上述 D/A 转换，必须使用解码网络。解码网络的主要形式有二进制权电阻解码网络和 T 形电阻解码网络两种。实际应用的 D/A 转换器多数采用 T 形电阻解码网络。由于它所采用的电阻阻值小，具有简单、直观、转换速度快、转换误差小等优点，因而本节仅介绍 T

形电阻解码网络的 D/A 转换。其结构原理图如图 10-1 所示。图中包括 1 个 4 位切换开关、4 路 R-2R 电阻网络、1 个运算放大器和 1 个比例电阻 R_F。

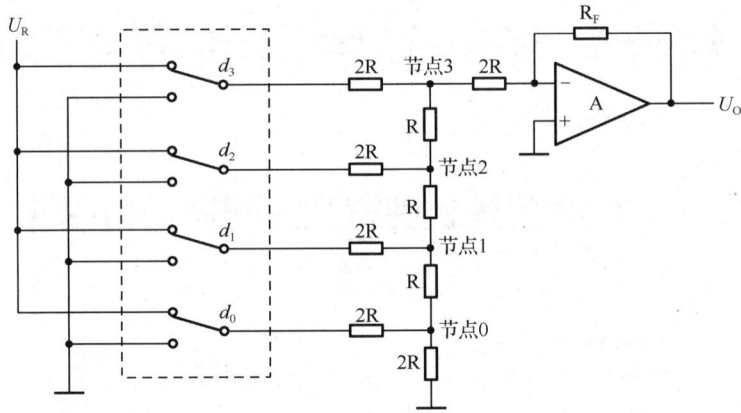

图 10-1　T 形电阻解码网络的 D/A 转换的结构原理图

整个 T 形电阻解码网络电路是由相同的电路环节组成的。每个电路环节都有 2 个电阻（R、2R）和 1 个开关，相当于二进制数的 1 位，开关由该位的代码控制。由于电阻接成 T 形，故称为"T 形电阻解码网络"。此电路采用了分流原理实现对输入位数字量的转换。图中无论从哪个节点向上或向下看，等效电阻都是 2R。从 $d_0 \sim d_3$ 看进去的等效输入电阻都是 3R，于是每个开关流入的电流 I 都可看作相等，即 $I=U_R/3R$。这样由开关 $d_0 \sim d_3$ 流入运算放大器的电流自上而下以 1/2 系数逐渐递减，依次为 $(1/2)I$、$(1/4)I$、$(1/8)I$、$(1/16)I$。设 $d_3d_2d_1d_0$ 为输入的二进制数字量，则输出的电压为

$$U_O = -R_F \sum I_i = -(R_F \times U_R/3R) \times (d_3 \times 2^{-1} + d_2 \times 2^{-2} + d_1 \times 2^{-3} + d_0 \times 2^{-4})$$

$$= -[(R_F \times U_R/3R) \times 2^{-4}] \times (d_3 \times 2^3 + d_2 \times 2^2 + d_1 \times 2^1 + d_0 \times 2^0)$$

式中，$d_0 \sim d_3$ 取值为 0 或 1。0 表示切换开关与地相连，1 表示切换开关与参考电压 U_R 接通，该位有电流输入。这就完成了由二进制数到模拟量电压信号的转换。由此式可以看出，U_R 和 U_O 的电压符号正好相反，即要使输出电压 U_O 为正，则 U_R 必须为负。由此式还可以看出，增加开关和权电阻的个数可以提高电压转换精度。

D/A 转换输出电压的大小不仅与二进制数码有关，还与运算放大器的反馈电阻 R_F、基准电压 U_R 有关。当 D/A 转换设置为满刻度值时，可以通过这两个参数调整电压的最大输出值。

2. D/A 转换的主要技术指标

（1）D/A 转换建立时间。

D/A 转换建立时间是描述转换速率高低的一个重要参数，是指当 D/A 转换器输入数字量为满刻度值（二进制各位全为 1）时，从输入加上模拟量电压到输出达到满刻度值或满刻度值的某一百分比（如 99%）所需的时间，也可称为输入 D/A 转换速率。不同类型的 D/A 转换建立时间大多是不同的，但一般均在几十纳秒到几百微秒的范围内。

（2）D/A 转换精度。

D/A 转换精度用于表明 D/A 转换的精确度，一般用误差大小来表示，以满刻度电压（满量程电压）U_{FS} 的百分数形式给出。例如，精确度为±0.1%指的是，最大误差为 U_{FS} 的±0.1%。

如果 U_{FS} 为 5V，则最大误差为±5mV。

（3）分辨率。

分辨率表示对输入的最小数字量的分辨能力，即当输入数字量最低位（LSB）产生一次变化时，所对应输出模拟量的变化量，而分辨率则与输入数字量的位数有关。如果输入数字量的位数为 n，则 D/A 转换器的分辨率为 2^{-n}。显然，在 D/A 转换器输出满量程电压相同的情况下，位数越多，分辨率就越高。通常以其二进制数码位数来表示分辨率。

注意：转换精度和分辨率是两个不同的概念。转换精度取决于构成转换器的各部件的误差和稳定性，而分辨率取决于转换器的位数。

10.1.2　DAC0832 的使用

1. DAC0832 的内部结构及功能

DAC0832 是采用 CMOS 工艺制作的 8 位单片梯形电阻式 D/A 转换器，片内带数据锁存器，电流输出型，输出电流持续时间为 1μs，其引脚排列图如图 10-2 所示。

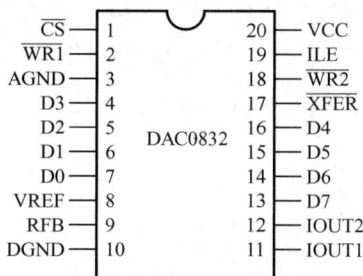

DAC0832 芯片
的使用微课视频

图 10-2　DAC0832 的引脚排列图

DAC0832 由一个 8 位 DAC 寄存器、一个 8 位输入锁存器、一个 8 位 D/A 转换器及逻辑控制电路组成。输入数据锁存器和 DAC 寄存器构成了两级缓存，可以实现多通道 D/A 的同步转换输出。由于 DAC0832 是电流型输出，因此应用时必须外接运算放大器使其成为电压型输出。

DAC0832 采用 20 脚的 DIP，各引脚的功能如下。

（1）D0～D7：8 位数据输入端，TTL 电平，有效时间长于 90ns。

（2）ILE：数据锁存允许控制信号输入端，高电平有效。

（3）$\overline{CS}$：片选信号输入端，低电平有效。

（4）$\overline{WR1}$：输入寄存器的写选通输入端，负脉冲有效（脉冲宽度应大于 500ns），当 $\overline{CS}$=0，ILE=1，有效时，D0～D7 状态被锁存到输入寄存器。

（5）$\overline{XFER}$：数据传输控制信号输入端，低电平有效。

（6）$\overline{WR2}$：DAC 寄存器写选通输入端，负脉冲（脉冲宽度应大于 500ns）有效，当 $\overline{XFER}$=0 且 $\overline{WR2}$ 有效时，输入寄存器的状态被传输到 DAC 寄存器中。

（7）IOUT1：电流输出端，当输入全为 1 时，I_{out1} 最大。

（8）IOUT2：电流输出端，其值和 I_{out1} 之和为一个常数。

（9）RFB：反馈电阻端，芯片内部此端与 IOUT1 之间已接有 1 个 15kΩ 的电阻。

（10）VCC：电源电压端，范围为+5～+15V。

（11）VREF：基准电压输入端，VREF 范围为 $-10 \sim +10V$，此端电压决定 D/A 输出电压的范围。如果 VREF 接 $+10V$，则输出电压范围为 $0 \sim -10V$，如果 VREF 接 $-5V$，则输出电压范围为 $0 \sim +5V$。

（12）AGND：模拟地，为模拟信号和基准电源的参考地。

（13）DGND：数字地，为工作电源地和数字逻辑地。两种地线最好在电源处一点共地。

2. DAC0832 的应用

DAC0832 与 51 单片机主要有三种基本的接口方式，即直通方式、单缓冲方式和双缓冲方式。

直通方式：使所有控制信号（$\overline{CS}$、$\overline{WR1}$、$\overline{WR2}$、$\overline{XFER}$）均有效，只适用于连续反馈控制线路。

单缓冲方式：适用于只有一路模拟量输出或几路模拟量非同步输出的情况。在这种方式下，将 2 级寄存器的控制信号并接，输入数据在控制信号的作用下，直接送入 DAC 寄存器，也可以采用把 $\overline{WR2}$、$\overline{XFER}$ 这两个信号固定接地的方法。

双缓冲方式：先控制 DAC0832 的数据锁存器以接收数据，然后控制 DAC0832 的 DAC 寄存器，通过这种方式可以实现多个 D/A 转换的同步输出。

10.1.3 DAC0832 正弦波信号发生器的任务实施

【设计要求】

通过 AT89C51 和 DAC0832 设计一个正弦波信号发生器，仿真图如图 10-3 所示。

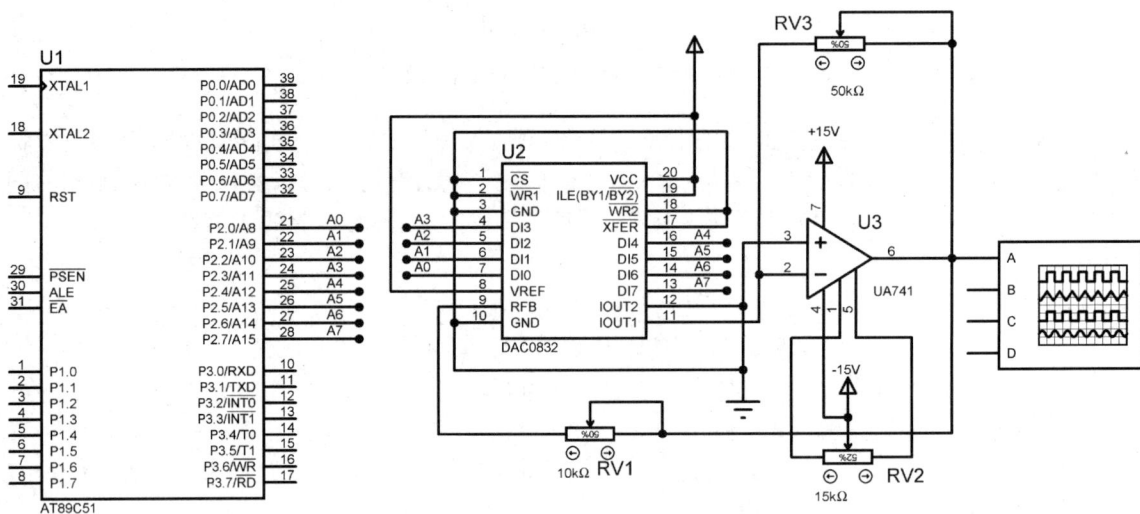

图 10-3 DAC0832 正弦波信号发生器仿真图

【任务分析】

将正弦波信号进行电压采样，一个周期采样点为 128 个，在程序中建立一个一维数组，数组长度为 128。可以通过单片机小工具"正弦波表生成器"计算出此一维数组的数据，如图 10-4 所示。在主程序设计中，主要按顺序在数组中查表，通过 P2 口输出。

图 10-4　正弦波表生成器产生正弦波数据

【实施步骤】

1. 添加元器件

打开 Proteus 仿真软件，按照表 10-1 添加元器件。

表 10-1　DAC0832 正弦波信号发生器的元器件清单

元器件名称	所属类	所属子类
AT89C51	Microprocessor ICs	8051 Family
UA741	Operational Amplifiers	Single
POT-HG	Resistors	Variable
DAC0832	Data Converters	D/A Converters

2. 绘制仿真图

元器件全部添加后，在 Proteus ISIS 的原理图编辑窗口中按图 10-3 绘制 DAC0832 正弦波信号发生器仿真图。

3. 编写程序

在 Keil μVision5 中编写程序，实现 DAC0832 正弦波信号发生器的效果，参考程序如下：

```
/*********************************************
程序名称：program10-1.c
程序功能：DAC0832 正弦波信号发生器程序
*********************************************/
#include<reg52.h>                        //加载头文件
/*********************************************
数据类型定义
*********************************************/
#define uchar unsigned char              //定义无符号字符型
/*********************************************
数组定义：正弦波信号产生数组，共 128 个点
*********************************************/
uchar code sin_dat[128]=
```

DAC0832 正弦波信号发生器参考程序

```
{
    0x7f,0x85,0x8b,0x91,0x97,0x9d,0xa3,0xa9,
    0xaf,0xb5,0xba,0xc0,0xc5,0xca,0xcf,0xd4,
    0xd8,0xdd,0xe1,0xe5,0xe8,0xeb,0xef,0xf1,
    0xf4,0xf6,0xf8,0xfa,0xfb,0xfc,0xfd,0xfd,
    0xfe,0xfd,0xfd,0xfc,0xfb,0xfa,0xf8,0xf6,
    0xf4,0xf1,0xef,0xeb,0xe8,0xe5,0xe1,0xdd,
    0xd8,0xd4,0xcf,0xca,0xc5,0xc0,0xba,0xb5,
    0xaf,0xa9,0xa3,0x9d,0x97,0x91,0x8b,0x85,
    0x7f,0x79,0x73,0x6d,0x67,0x61,0x5b,0x55,
    0x4f,0x49,0x44,0x3e,0x39,0x34,0x2f,0x2a,
    0x26,0x21,0x1d,0x19,0x16,0x13,0x0f,0x0d,
    0x0a,0x08,0x06,0x04,0x03,0x02,0x01,0x01,
    0x00,0x01,0x01,0x02,0x03,0x04,0x06,0x08,
    0x0a,0x0d,0x0f,0x13,0x16,0x19,0x1d,0x21,
    0x26,0x2a,0x2f,0x34,0x39,0x3e,0x44,0x49,
    0x4f,0x55,0x5b,0x61,0x67,0x6d,0x73,0x79,
};
/************************************************************************
函数名：void main()
功能描述：主函数
*************************************************************************/
void main()
{
    uchar i;
    while(1)
    {
        for(i=0;i<128;i++)                    //以下为DAC0832驱动程序
            P2=sin_dat[i];                    //P2口输出数据
    }
}
```

4. 系统仿真

当 Keil C51 编译成功后，会自动产生 HEX 文件，接着打开之前绘制的 Proteus 仿真图，双击 AT89C51，弹出"Edit Component"对话框，单击"Program File"中的文件夹按钮，在弹出的"Select File Name"对话框中，选择之前编译生成的 HEX 文件，单击"打开"按钮，返回"Edit Component"对话框，单击"OK"按钮，即可装入 HEX 文件。

DAC0832 正弦波信号
发生器仿真效果视频

接着单击 Proteus ISIS 编辑界面左下角的运行按钮▶，即可通过示波器观察是否能够实现 DAC0832 正弦波信号发生器的显示效果，如图 10-5 所示。

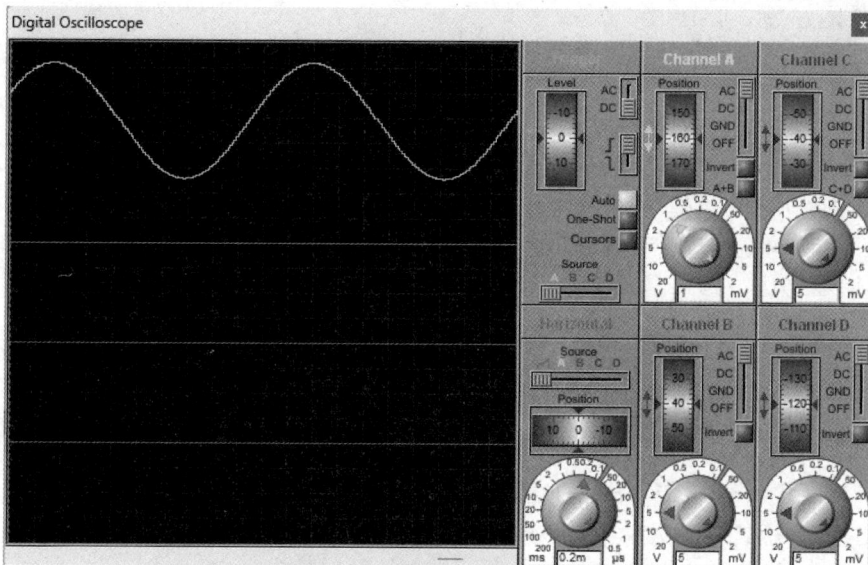

图 10-5　示波器仿真显示效果图

任务 10.2　TLC5615 三角波信号发生器的仿真设计

学习目标

【知识目标】

（1）了解并掌握 TLC5615 的基本原理和使用方法。

（2）了解并掌握 TLC5615 三角波信号发生器的实现方法。

【技能目标】

（1）了解并掌握单片机仿真软件 Proteus 的使用方法。

（2）了解并掌握单片机编译软件 Keil C51 的使用方法。

（3）通过 TLC5615 三角波信号发生器的仿真设计初步了解并掌握单片机项目的开发步骤。

【思政目标】

在 TLC5615 三角波信号发生器的仿真设计中，要求学生严格按照科学方法进行设计和测试，准确记录和分析数据，提高三角波信号的线性度。同时，鼓励学生勇于面对问题、敢于挑战困难，培养坚韧不拔、实事求是的科学态度。

TLC5615 三角波信号
发生器导学材料

10.2.1　TLC5615 的使用

1. TLC5615 的特点

TLC5615 芯片的
使用微课视频

TLC5615 是一个串行 10 位 D/A 转换器，其性能比早期电流型输出的 D/A 转换器要好。只需通过 3 根串行总线就可以完成 10 位数据的串行输入，易于和工业标准的微处理器或微控制器相接，适用于数字失调与增益调整及工业控制场合。

TLC5615 的特点如下。

（1）单 5V 电源工作。

（2）3 线串行接口。

（3）高阻抗基准输入端。

（4）输出最大电压为 2 倍基准输入电压。

（5）上电时内部自动复位。

（6）低功耗，最大功耗为 1.75mW。

（7）转换速率快，更新率为 1.21MHz。

（8）具有单、双极性输出。

（9）可编程的 MSB 或 LSB 前导。

（10）可编程的输出数据长度。

2. TLC5615 的引脚排列图和内部结构框图

TLC5615 的引脚排列图如图 10-6 所示。各引脚功能如下。

（1）DIN：串行二进制数输入端。

（2）SCLK：串行时钟输入端。

（3）CS：片选端，低电平有效。

（4）DOUT：用于级联的串行数据输出端。

（5）AGND：模拟地。

（6）REFIN：基准电压输入端，大小为 2V～（VCC-2V），典型值为 2.048V。

（7）OUT：模拟电压输出端。

（8）VCC：电源端，一般接+5V。

TLC5615 的内部结构框图如图 10-7 所示，主要由以下几部分组成。

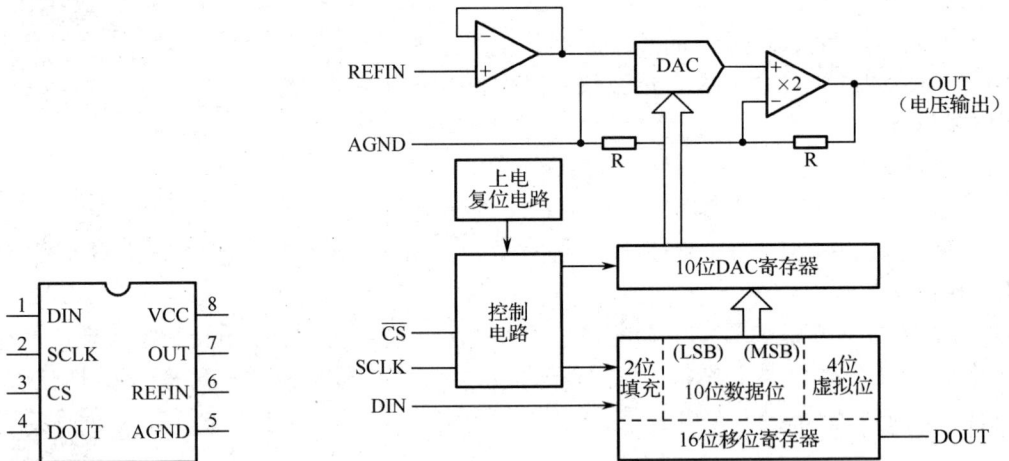

图 10-6　TLC5615 的引脚排列图

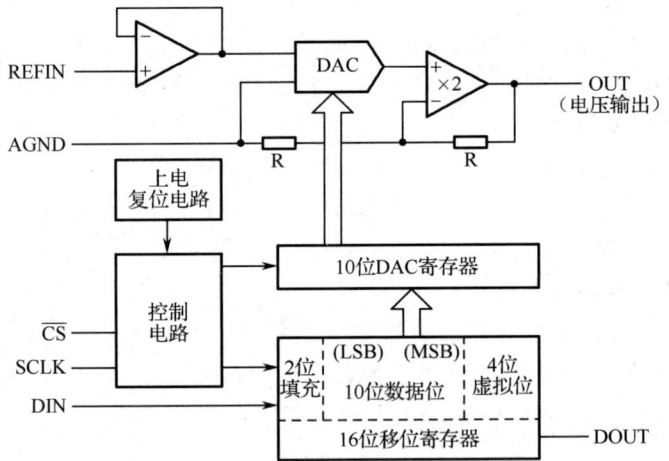

图 10-7　TLC5615 的内部结构框图

（1）10 位 DAC 电路。

（2）16 位移位寄存器，接收串行移入的二进制数，并且有一个级联的数据输出端 DOUT。

（3）并行输入/输出的 10 位 DAC 寄存器，为 10 位 DAC 电路提供待转换的二进制数。

（4）电压跟随器为 REFIN 端提供很高的输入阻抗，大约 10MΩ。

（5）×2 电路提供最大值为 REFIN 端 2 倍的输出。

（6）上电复位电路和控制电路。

3．TLC5615 的使用方法

TLC5615 通过固定增益为 2 的运放缓冲电阻串网络，把 10 位数据字转换为模拟电压电平。上电时，内部电路把 DAC 寄存器复位为 0。其输出具有与基准输入相同的极性，表达式为

$$U_O = \frac{2 \times REF \times CODE}{1024}$$

（1）数据输入。

由于 DAC 寄存器是 12 位寄存器，所以在 10 位数据字中必须写入数值为 0 的 2 个低于 LSB（D0）的位（次最低有效位）。

（2）D/A 输出。

输出缓冲器具有满电源电压幅度输出，具有短路保护并能驱动 100pF 负载电容的 2kΩ 负载。

（3）外部基准。

基准电压输入经过缓冲，使得 DAC 输入电阻与代码无关。因此 REFIN 输入电阻为 10MΩ，REFIN 输入电容的典型值为 5pF，它们与输入代码无关。基准电压取决于 DAC 的满度输出。

（4）逻辑接口。

逻辑输入端可以使用 TTL 或 CMOS 逻辑电平。但是用满电源电压幅度，CMOS 逻辑可得到最低的功耗。当使用 TTL 逻辑电平时，功率需求增加约两倍。

（5）菊花链接器件。

假如时序关系合适，则可以通过在一个链路中把一个器件的 DOUT 端连接到下一个器件的 DIN 端实现 DAC 的菊花链接（级联）。DIN 处的数据延迟 16 个时钟周期加一个时钟宽度后出现在 DOUT 端。DOUT 是低功率的推拉输出电路。当 CS 端为低电平时，DOUT 端在 SCLK 端下降沿变化；当 CS 端为高电平时，DOUT 端保持在最近数据位的值并不进入高阻态。

（6）数据格式。

当 CS 端为低电平时，输入数据读入 16 位移位寄存器（由时钟同步，最高有效位在前）。SCLK 端输入的上升沿把数据移入输入寄存器。接着，CS 端的上升沿把数据传输至 DAC 寄存器。当 CS 端为高电平时，输入数据不能由时钟同步送入输入寄存器。所有 CS 端的跳变应当发生在 SCLK 端输入为低电平时。

如果不使用菊花链接功能，那么可以使用 MSB 在前的 12 位输入数据序列：

D9	D8	D7	D6	D5	D4	D3	D2	D1	D0	0	0

如果使用菊花链接功能，那么可以传输 4 个高虚拟位在前的 16 位输入数据序列：

4 个高虚拟位		10 位数据位		0	0

来自 DOUT 端的数据需要输入时钟 16 个下降沿，因此，需要额外的时钟宽度。当菊花链接多个 TLC5615 时，因为数据传输需要 16 个输入时钟周期加上一个额外的输入时钟下降沿使数据在 DOUT 端输出，所以，数据需要 4 个高虚拟位。为了提供与 12 位数据转换器传输的硬件与软件兼容性，两个额外位总是需要的。

（7）系统稳定性及功耗。

为了更好地使用 TLC5615，建议使用分离的模拟地和数字地平面来提高系统性能。设计两个地平面时，应当在低阻抗处将模拟地与数字地连接在一起。通过把器件的 AGND 端连接到系统模拟地平面（该平面能确保模拟地电流流动良好且地平面上的电压降可以忽略），可以实现最佳的接地连接。

VCC 和 AGND 之间应连接一个 0.1μF 的陶瓷旁路电容，且应当用短引线安装在尽可能靠近器件的地方。

当系统不使用 D/A 转换器时，将 DAC 寄存器设置为全 0，可以使基准电阻阵列和输出负载的功耗降为最低。

10.2.2　TLC5615 三角波信号发生器的任务实施

【设计要求】

使用 AT89C51 和 TLC5615 设计一个周期为 30ms 的三角波信号发生器，仿真图如图 10-8 所示。

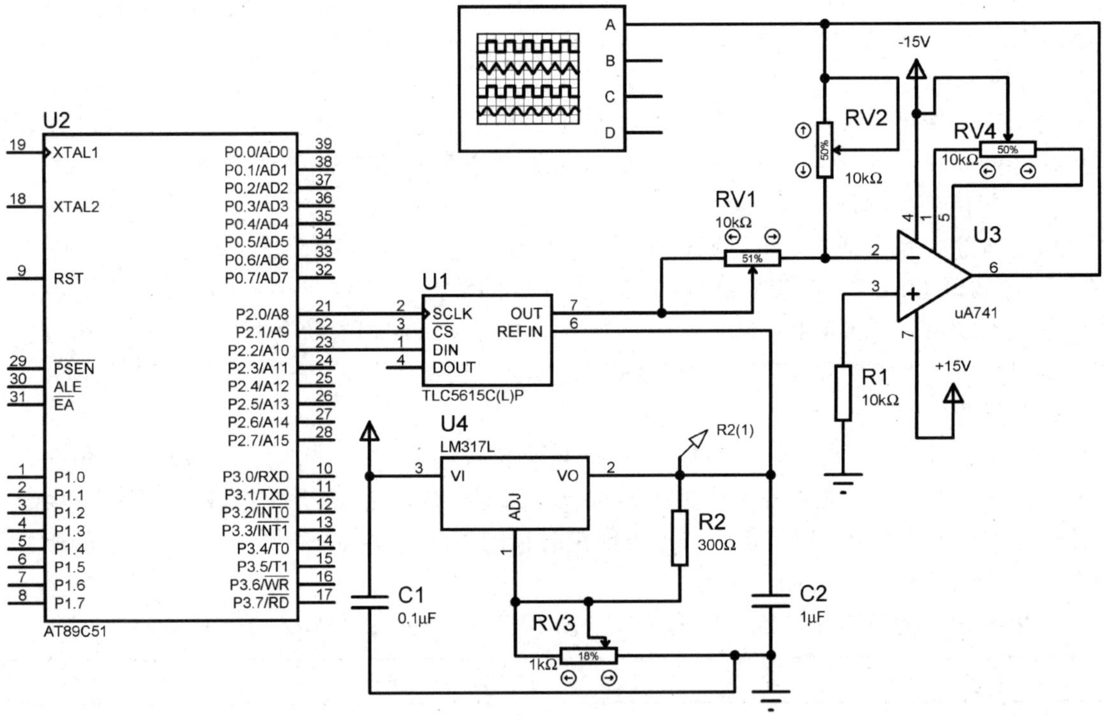

图 10-8　TLC5615 三角波信号发生器仿真图

【任务分析】

在本任务中，设定 TLC5615 的基准电压为 2V，因此可以通过 LM317 三端可调稳压器将电源电压（5V）稳成 2V 实现，在 TLC5615 的输出端连接一个反相比例运算放大电路，可以调节输出信号的幅度。

【实施步骤】

1. 添加元器件

打开 Proteus 仿真软件，按照表 10-2 添加元器件。注意：用 Proteus 仿真软件绘制单片机仿真图时，可以省略振荡电路和复位电路。

表 10-2　TLC5615 三角波信号发生器的元器件清单

元器件名称	所属类	所属子类
AT89C51	Microprocessor ICs	8051 Family
RES	Resistors	Generic
AT89C51	Microprocessor ICs	8051 Family
UA741	Operational Amplifiers	Single
POT-HG	Resistors	Variable
TLC5615	Data Converters	D/A Converters
LM317L	Analog ICs	Regulators
CAP	Capacitors	Generic

2. 绘制仿真图

元器件全部添加后，在 Proteus ISIS 的原理图编辑窗口中按图 10-8 绘制 TLC 三角波信号发生器仿真图。

TLC5615 三角波信号发生器参考程序

3. 编写程序

在 Keil μVision5 中编写程序，实现 TLC5615 三角波信号发生器的效果，参考程序如下：

```
/*******************************************************************
程序名称：program10-2.c
程序功能：TLC5615 三角波信号发生器程序
*******************************************************************/
#include<reg52.h>                        //加载头文件
/*******************************************************************
数据类型定义
*******************************************************************/
#define uchar unsigned char              //定义无符号字符型
#define uint unsigned int                //定义无符号整型
/*******************************************************************
单片机引脚定义
*******************************************************************/
sbit CK = P2^0;                          //时钟信号接口定义
sbit CS = P2^1;                          //片选信号接口定义
sbit DIN = P2^2;                         //数据接口定义
/*******************************************************************
全局变量定义
*******************************************************************/
uchar LCD1602_DATA;                      //LCD1602 待写入数据存储单元
```

```
uchar ADC_DATA=0;                                    //ADC0831 数据存储单元
uchar DISPLAY_DATA[4]={0};                            //显示存储单元
/****************************************************************
函数名：tlc5615(uint data)
功能描述：TLC5615 驱动程序
输入参数：dat
说明：dat 是进行 D/A 转换的数据
****************************************************************/
void tlc5615(uint dat)
{
    uchar i;
    CS = 0;
    CK = 0;                                          //启动 TLC5615
    for (i=0;i<12;i++)
    {
        DIN =dat&0x200;                              //发送数据
        CK = 1;
        dat<<= 1;
        CK=0;                                        //产生时钟信号
    }
    CS = 1;                                          //转换数据发送完毕后关闭 TLC5615
}
/****************************************************************
函数名：void main()
功能描述：主函数
****************************************************************/
void main()
{
    uint i;
    while(1)
    {
        for(i=1023;i>10;i=i-20)                      //产生三角波信号
            tlc5615(i);                             //调用 TLC5615 驱动程序
        for(i=0;i<1023;i=i+20)                       //产生三角波信号
            tlc5615(i);                             //调用 TLC5615 驱动程序
    }
}
```

4. 系统仿真

当 Keil C51 编译成功后，会自动产生 HEX 文件，接着打开之前绘制的 Proteus 仿真图，双击 AT89C51，弹出"Edit Component"对话框，单击"Program File"中的文件夹按钮，在弹出的"Select File Name"对话框中，选择之前编译生成的 HEX 文件，单击"打开"按钮，返回"Edit Component"对话框，单击"OK"按钮，即可装入 HEX 文件。

TLC5615 三角波信号
发生器仿真效果视频

接着单击 Proteus ISIS 编辑界面左下角的运行按钮 ▶，即可通过示波器观察是否能够实现 TLC5615 三角波信号发生器的显示效果，如图 10-9 所示。

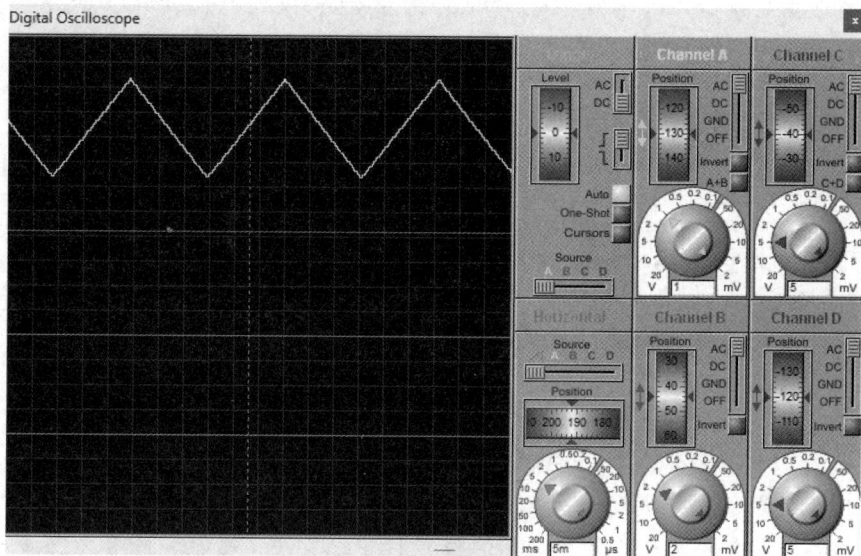

图 10-9　示波器仿真显示效果图

素养小课堂

单片机应用系统设计的步骤

单片机应用系统设计包括硬件设计和软件设计两部分。一般来说，应用系统所要完成的任务不同，相应的硬件和软件也就不同。硬件软件化是提高系统性价比的有效方法，尽量减少硬件成本，多用软件实现相同的功能，这样也可以大大提高系统的可靠性。

为了保证系统能够可靠地工作，在软/硬件设计中，还要考虑系统的抗干扰设计。虽然单片机的硬件选型不尽相同，软件编写也千差万别，但系统的开发步骤和方法是基本一致的，一般分为总体设计、硬件设计、软件设计、仿真调试、资料整理五个阶段。单片机应用系统的开发流程如图 10-10 所示。

1. 总体设计

总体设计包括需求分析和方案论证等，是单片机应用系统设计工作的开始和基础。只有经过深入细致的需求分析和周密科学的方案论证，才能使系统设计工作顺利完成。先确立任务，再进行单片机型号选择，最后进行方案论证。

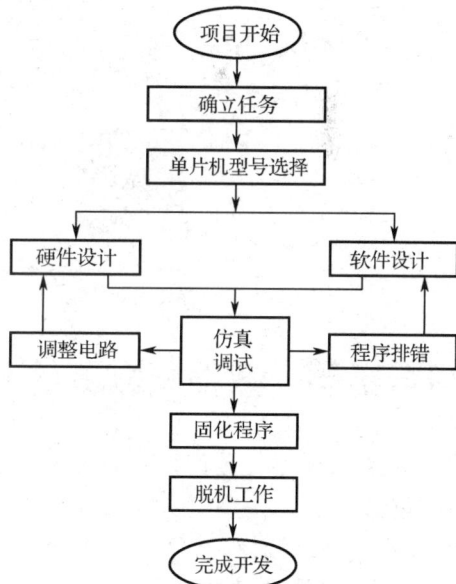

图 10-10　单片机应用系统的开发流程

2. 硬件设计

根据总体设计中确立的功能特性要求，选择单片机的型号(单片机最初的型号选择很重要，原则上是选择高性价比的单片机)、所需外围扩展芯片、存储器、I/O 电路、驱动电路，可能还有 A/D 和 D/A 转换电路及其他模拟电路，设计出单片机应用系统的电路原理图。

3. 软件设计

进行系统资源分配，设计程序结构及程序流程图，编写程序。

4. 仿真调试

硬件和软件设计完成后，一般不能按预计的任务正常工作，需要查错和调试。调试时，应将硬件和软件分成几部分，逐个进行调试，然后进行联调，并进行性能测定。

5. 资料整理

资料不仅是设计工作的结果，还是以后使用、维修及进一步设计的依据。资料应包括任务描述、设计思路及方案论证、性能测定及软件资料（流程图、函数使用说明、参考程序）和硬件资料（电路原理图、线路板图、注意事项等）。

从总体上来看，设计任务分为硬件设计和软件设计两部分，两者缺一不可。硬件设计的绝大部分工作量是在前期完成的，后期只需做一些修改；软件设计任务贯彻始终，到中后期基本上都是软件设计任务。在单片机应用系统设计中，软件、硬件和抗干扰设计是紧密相关、不可分离的。设计者应根据实际情况，合理地安排软/硬件的比例，选取最佳的设计方案，使系统具有最佳的性价比。

课后任务

1. 如图 10-3 所示，通过 AT89C51 和 DAC0832 设计一个三角波信号发生器。显示效果参见二维码。

2. 如图 10-8 所示，通过 AT89C51 和 TLC5615 设计一个正弦波信号发生器。显示效果参见二维码。

课后任务 1
仿真效果视频

课后任务 2
仿真效果视频

知识拓展　数据获取器件 PCF8591

PCF8591 是一款单片集成、单独供电、低功耗、8 位 CMOS 数据获取器件，其功能包括多路模拟输入、内置跟踪保持、8 位 A/D 转换和 8 位 D/A 转换。它既可以进行 A/D 转换，又可以进行 D/A 转换，进行 A/D 转换时为逐次逼近式。PCF8591 的地址、控制和数据信号都通过 I^2C 总线以串行的方式进行传输。PCF8591 的最高转换速率由 I^2C 总线的最高速率决定。

PCF8591 为 16 引脚，采用 SOP 或 DIP，其引脚排列图如图 10-11 所示。PCF8591 是一个单电源低功耗的 8 位 CMOS 数据获取器件，具有 4 路模拟输入、1 路模拟输出、1 个串行 I^2C

总线接口，用来与单片机通信。3 个地址引脚 A0、A1、A2，用于编程硬件地址，允许最多 8 个器件连接到 I^2C 总线，而不需要额外的片选电路。器件的地址、控制及数据都通过 I^2C 总线来传输。

在 PCF8591 内部的可编程功能控制字有两个，一个为地址选择字，另一个为转换控制字。PCF8591 采用典型的 I^2C 总线接口的器件寻址方法，即总线地址由器件地址、引脚地址和方向位组成，如图 10-12 所示。NXP 公司规定 A/D 转换器高 4 位地址为 1001，低 3 位地址为引脚地址 A0A1A2，由硬件电路决定。所以，I^2C 总线中最多可接 $2^3=8$ 个具有 I^2C 总线接口

图 10-11 PCF8591 的引脚排列图

的 A/D 转换器，地址的最后一位为方向位 R/$\overline{W}$，当主控器件对 A/D 转换器进行读操作时为 1，进行写操作时为 0。总线操作时，地址选择字为主控器件发送的第 1 个字节。

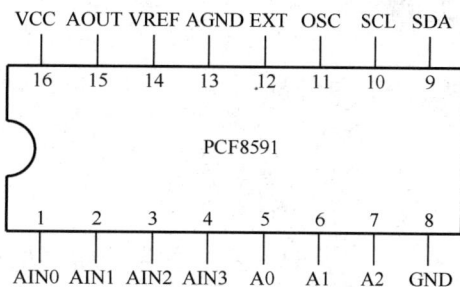

PCF8591 的转换控制字存放在控制寄存器中，用于实现器件的各种功能。总线操作时，转换控制字为主控器件发送的第 2 个字节，转换控制字的格式如图 10-13 所示。

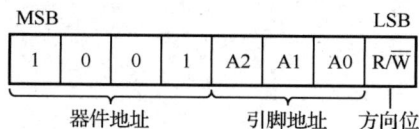

图 10-12 地址选择字的格式

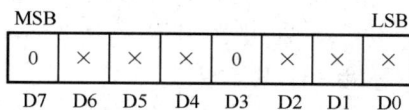

图 10-13 转换控制字的格式

转换控制字各位的功能如下。

（1）D0D1：通道选择位，00 表示通道 0，01 表示通道 1，10 表示通道 2，11 表示通道 3。

（2）D2：自动增量允许位，为 1 时，每次对一个通道进行转换后，自动切换到下一个通道进行转换；为 0 时，不自动进行转换，可通过软件修改进行转换。

（3）D3：特征位，固定为 0。

（4）D4D5：模拟量输入方式选择位。00 表示输入方式 0，4 路单端输入；01 表示输入方式 1，3 路差分输入；10 表示输入方式 2，2 路单端输入，1 路差分输入；11 表示输入方式 3，2 路差分输入。

（5）D6：模拟输出允许位，A/D 转换时设置为 0，D/A 转换时设置为 1。

（6）D7：特征位，固定为 0。

D/A 转换器是 PCF8591 的关键单元，通过它进行 D/A 转换时执行 I^2C 总线的写入操作。PCF8591 的 D/A 转换数据格式如图 10-14 所示。其中，S 为 I^2C 总线启动信号位，第 1 个字节 SLAW 为主控器件（单片机）发送的 PCF8591 地址选择字，第 2 个字节 CONBYT 为主控器件发送的 PCF8591 转换控制字，data 1～data n 为待转换的二进制数，A 为 1 字节传输完后由 PCF8591 产生的应答信号位，P 为 I^2C 总线停止信号位。

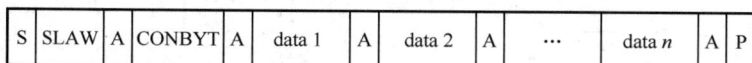

S	SLAW	A	CONBYT	A	data 1	A	data 2	A	…	data n	A	P

图 10-14 PCF8591 的 D/A 转换数据格式

习题

一、单选题

1. DAC0832 是（　　）芯片。

A. 8 位 D/A 转换器　　　　　　　　　　　B. 16 位 D/A 转换器

C. 8 位 A/D 转换器　　　　　　　　　　　D. 16 位 A/D 转换器

2. 多路 D/A 转换系统中为实现多路模拟信号同步输出，常采用（　　）。

A. 直通方式　　　　B. 单缓冲方式　　　　C. 双缓冲方式　　　　D. 均可

3. D/A 转换可以实现（　　）功能。

A. 模拟信号转换为数字信号　　　　　　　B. 数字信号转换为模拟信号

C. 编码　　　　　　　　　　　　　　　　D. 数据选择

4. 若采用 8 位 D/A 转换器，输入数字量为 0x70，满量程为 5V，则输出电压为（　　）。

A. 1.19V　　　　　B. 2.19V　　　　　C. 3.19V　　　　　D. 4.19V

5. 关于 TLC5615，以下说法中错误的是（　　）。

A. 采用单 5V 电源工作

B. 采用 3 线串行接口

C. D/A 转换器输出最大电压为 3 倍基准输入电压

D. 具有单、双极性输出

二、设计题

1. 请采用单片机和 PCF8591 实现数字电压表的仿真设计，采用
LCD1602 进行显示，其电路图如图 10-15 所示，显示效果参见二维码。

设计题 1 仿真效果视频

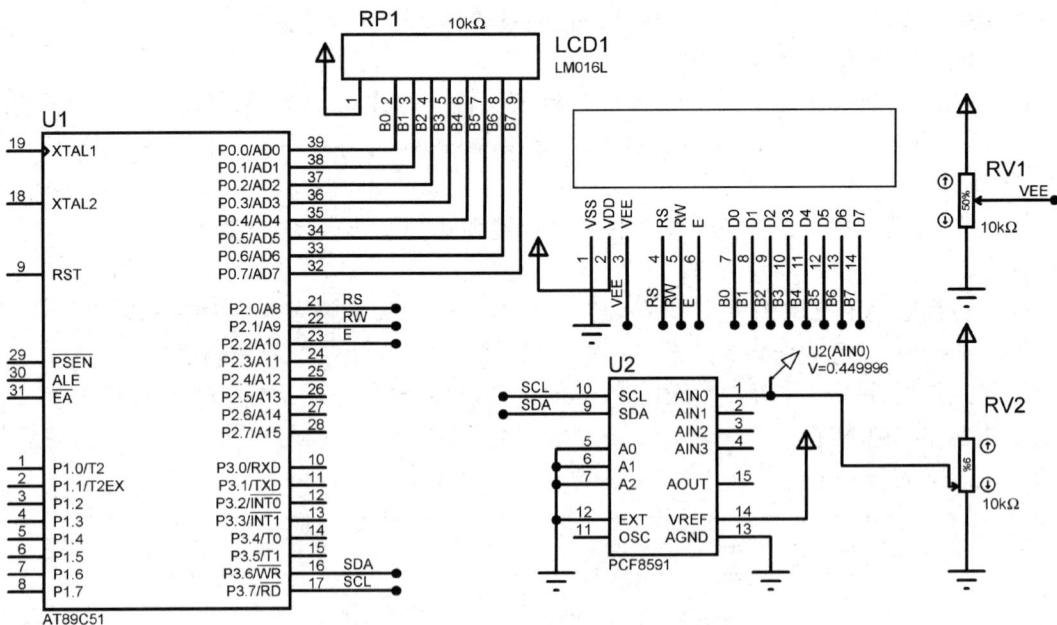

图 10-15　PCF8591 数字电压表电路图

2．请采用单片机和 PCF8591 实现正弦波信号发生器的仿真设计，其电路图如图 10-16 所示，显示效果参见二维码。

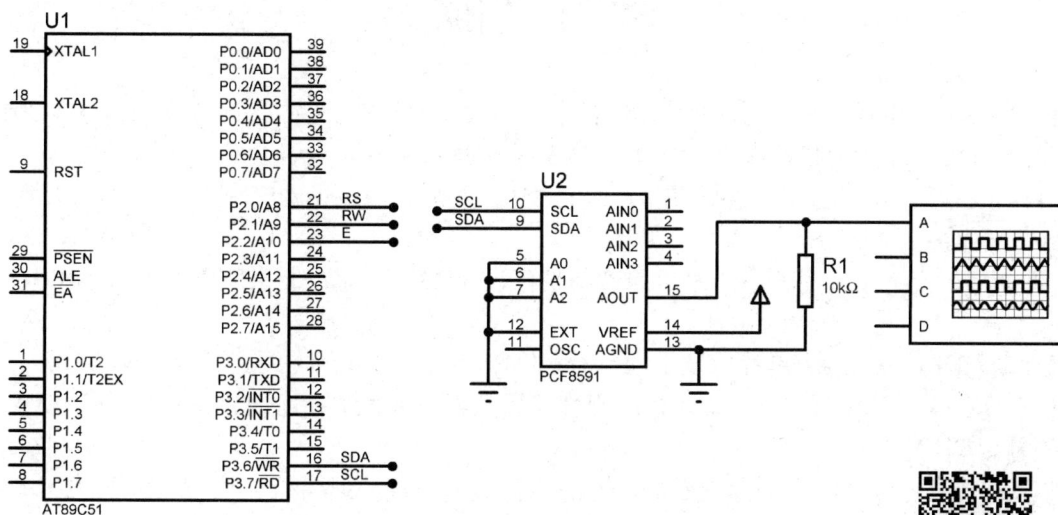

图 10-16　PCF8591 正弦波信号发生器电路图

设计题 2 仿真效果视频

参考文献

[1] 王静霞. 单片机基础与应用（C 语言版）（第 2 版）[M]. 北京：高等教育出版社，2019.

[2] 李建兰. 单片机原理及接口技术（第 3 版）[M]. 北京：电子工业出版社，2023.

[3] 卓书芳，何用辉. 单片机应用技术项目教程（基于 Proteus 的 C 语言版）[M]. 北京：机械工业出版社，2022.

[4] 叶钢，李三波，张莉. 单片机原理与仿真设计[M]. 北京：北京航空航天大学出版社，2009.

[5] 徐国华，刘春艳. 单片机技术项目教程[M]. 北京：北京师范大学出版社，2018.

[6] 马忠梅，等. 单片机的 C 语言应用程序设计（第 6 版）[M]. 北京：北京航空航天大学出版社，2017.

[7] 黄锡泉，何用辉. 单片机技术及应用：基于 Proteus 的汇编和 C 语言版[M]. 北京：机械工业出版社，2014.

[8] 张毅刚. 单片机原理与应用设计：C51 编程+Proteus 仿真（第 2 版）[M]. 北京：电子工业出版社，2015.

[9] 孟凤果. 单片机应用技术项目式教程（C 语言版）[M]. 北京：机械工业出版社，2017.

[10] 王静霞. 单片机应用技术（经典项目化案例式新形态活页教材）（第 5 版）[M]. 北京：电子工业出版社，2023.

[11] 陈海松. 单片机应用技能项目化教程（第 2 版）[M]. 北京：电子工业出版社，2012.

[12] 李全利. 单片机原理及应用技术——基于 C51 的 Proteus 仿真及实板案例（第 4 版）[M]. 北京：高等教育出版社，2014.

[13] 韦龙新. 单片机技术与项目训练 [M]. 北京：电子工业出版社，2023.

[14] 高玉芹. 单片机原理与应用及 C51 编程技术 [M]. 北京：机械工业出版社，2011.

[15] 刘波. 51 单片机应用开发典型范例：基于 PROTEUS 仿真 [M]. 北京：电子工业出版社，2016.

反侵权盗版声明

电子工业出版社依法对本作品享有专有出版权。任何未经权利人书面许可，复制、销售或通过信息网络传播本作品的行为，歪曲、篡改、剽窃本作品的行为，均违反《中华人民共和国著作权法》，其行为人应承担相应的民事责任和行政责任，构成犯罪的，将被依法追究刑事责任。

为了维护市场秩序，保护权利人的合法权益，我社将依法查处和打击侵权盗版的单位和个人。欢迎社会各界人士积极举报侵权盗版行为，本社将奖励举报有功人员，并保证举报人的信息不被泄露。

举报电话：（010）88254396；（010）88258888

传　　真：（010）88254397

E-mail： dbqq@phei.com.cn

通信地址：北京市海淀区万寿路 173 信箱

　　　　　电子工业出版社总编办公室

邮　　编：100036